PASSIONS FOR BIRDS

Passions for Birds

Science, Sentiment, and Sport

SEAN NIXON

McGill-Queen's University Press
Montreal & Kingston • London • Chicago

ISBN 978-0-2280-1045-6 (cloth)
ISBN 978-0-2280-1046-3 (ePDF)
ISBN 978-0-2280-1047-0 (ePUB)

Legal deposit second quarter 2022
Bibliothèque nationale du Québec

Printed in Canada on acid-free paper that is 100% ancient forest free (100% post-consumer recycled), processed chlorine free

---

Library and Archives Canada Cataloguing in Publication

Title: Passions for birds : science, sentiment, and sport / Sean Nixon.
Names: Nixon, Sean, 1966– author.
Description: Includes bibliographical references and index.
Identifiers: Canadiana (print) 20220149208 | Canadiana (ebook) 20220149313 | ISBN 9780228010456 (cloth) | ISBN 9780228010463 (ePDF) | ISBN 9780228010470 (ePUB)
Subjects: LCSH: Birds—Social aspects. | LCSH: Human–animal relationships.
Classification: LCC QL673 .N59 2022 | DDC 598—dc23

This book was typeset in 10.5/13 New Baskerville ITC Pro.
Copy-editing and composition by T&T Productions Ltd, London.

Contents

Figures

Acknowledgements

The arguments developed in this book have evolved over many years and have benefitted from being aired at various seminars and conferences. I am especially grateful to audiences and fellow speakers at the 'J.A. Baker, Life and Legacy Conference' (University of Essex, July 2016), 'Understanding Material Loss in Time and Space' (a 'Past and Present' conference, University of Birmingham, February 2017), the Design History Conference (University of Oslo, September 2017), the Royal College of Art/Victoria & Albert Museum History of Design seminar series (January 2018), the Royal Geographical Society Annual Conference (Cardiff, August 2018), the Social History Society Annual Conference (University of Lincoln, 2018), and the 'Winged Geographies' conference (Cambridge University, April 2021).

Given the historical focus of my account, libraries and archives have figured centrally in the research. Elizabeth George was supportive and knowledgeable in helping me to negotiate the RSPB's Library, while Adam Rowlands kindly allowed me access to the Minsmere Warden's reports. David Badger helped with the RSPB's film library. Staff at the British Library Reading Rooms; the BBC's Written Archives, Caversham Park; the Linnaean Society Library; the Beinecke Library, Yale University; the Albert Sloman Library, University of Essex; the Alexander Library Archive and Manuscript Collection; the Bodleian Library, University of Oxford; the BTO Library, Thetford; and the National Archives, Kew, all helped me to access their important collections.

Many colleagues and friends have shaped my thinking and I am particularly grateful to Shaul Bar-Haim, Ted Benton, James Canton,

Jonathan Dean, Paul du Gay, Michael Lane, Jeremy Mynott, Jules Pretty, Michael Roper, and Robin West. Michael Roper, Ted Benton, and Michael Lane read the majority of the manuscript and I am particularly grateful to them for their careful and generous comments. The two anonymous reviewers of the manuscript also gave excellent comments, for which I am grateful. Richard Baggaley has been an exemplary editor to work with, and his belief in and enthusiasm for the book has been much appreciated.

Claire Nixon has lived with this project from its inception, offering wise counsel, love, and support. She has made its completion possible. The book is dedicated to her, with love.

PASSIONS FOR BIRDS

Introduction

This book is about human passions for wild birds. It explores the relationships people forged with them in bird conservation and recreational birdwatching, in the field sports of falconry and wildfowling, and within the residual cultures of harvesting wild birds for food. During the central decades of the twentieth century, these were all culturally significant ways in which people were drawn close to wild birds in Britain, Western Europe, and the wider North Atlantic world. Bringing these different practices together is central to my argument. Doing so illuminates the shifting relations between people and wild birds across a key period of twentieth-century history and reveals the dialogue, cultural exchanges, and conflicts that animated the relationships between the different bird-centred practices.

Understanding the relationships between birdwatching, bird conservation, field sports, and harvesting cultures matters because the connections between them were a live and active dimension of how each practice ordered human–avian relations. All of the practices shared deep feelings for wild birds and sought technological and material means of bringing birds closer to people. These emotional, subjective, and material forms of attachment are at the heart of the chapters that follow. I show how modern birdwatchers, wildfowlers, falconers, and bird harvesters often shared remarkably similar feelings and passions for birds. A range of material objects and devices – from hides, traps, nets, and ropes to binoculars – additionally linked these practices and their relations with birds. It is these commonalities of feeling and material connection, as much as an exploration of the divergent styling of human–avian relations across these practices, that my account sets out to explore.

Driving my argument is a particular understanding of cultural change. This foregrounds processes of transgression, translation, and hybridization in accounting for the emergence of new social practices, rather than thinking of change as emerging through the substitution of the old by the new or in the effecting of clean breaks between historical moments and practices.[1] If social and cultural change proceeds, within this approach, through mixing, translating, and reworking, it also generates its own forms of ambivalence and the strange afterlife of old practices.

The new birdwatching promoted by its able advocates like Julian Huxley, James Fisher, Roger Tory Peterson, and others did not, from this perspective, break completely with the older natural history or with the bird-centred field sports and bird harvesting traditions. Rather, as I show, even as it emphasized its modernity and its attention upon the living bird, the new ornithology reworked and translated elements of these other relationships with wild birds, transforming them into something different. Thus, for example, the collecting imperatives of the old natural history were taken up and reworked by the new birdwatcher naturalists.

At the same time, watching birds retained, for many, a strong sporting element, with the associated quest for trophies and the excitement of the hunt. As Roger Tory Peterson confessed, the 'game of birding' was a 'sport' as much it was indebted to the science of bird study.[2] Similarly, techniques for getting close to wild birds within sport and harvesting practices, and their respective knowledge of bird behaviour and environmental needs, were incorporated into modern birdwatching. In this sense, the new birdwatching remained ambivalent about the older and established cultures of nature, simultaneously absorbing and rejecting elements of these other ways of encountering wild birds.

Techniques and approaches could also move in the other direction. Field sports and bird harvesting often mixed an emphasis on their antiquity and an appeal to a customary consciousness and tradition, while modernizing their practices and presenting themselves as pro-conservation activities by drawing upon the language and practices of conservation-minded birdwatching.

In focusing on the productive antagonisms and translations that shaped the relations between birdwatching and the other bird-centred practices, the book challenges two dominant themes that have shaped histories of ornithology, conservation, and human–animal relations more broadly. These have, firstly, tended to see the emergence of humanitarian, welfare-based, and conservationist attitudes to wild animals as superseding older ways of ordering human–animal relations. Certainly, it has repeatedly been argued that for the majority of urban and suburban populations in the West, human relationships with wild animals came to be expressed principally through leisure and recreation; lived through practices of observation rather than killing for sport or food; and rooted in an ethic of conservation and welfare rather than exploitation.[3] Studies of ornithology and birdwatching in the United States and Britain have tended to follow this narrative. For example, in their rich histories of US and UK professional and amateur ornithology, Bircham, Burrow, and Dunlop have charted the break with the old natural history and its traditions of shooting and collecting.[4] In their studies, modern ornithology and recreational birdwatching appear as the successors to these older practices, replacing the earlier forms of attention upon wild birds. If this progressive narrative of the 'civilizing' of relations with wild birds undoubtedly catches the broad direction of social and cultural change, it nonetheless underplays the persistence of bloodier relations with wild animals. It also assumes a model of social progress that, as I have argued, misses how elements of the old continued and were reworked within the new.

The second problem with much of the historical literature is its tendency to segregate the histories of field sports, natural history, and conservation. Thus John Sheail, for example, in his fine-grained accounts of nature conservation in Britain, focuses on the relationship between policymakers, conservation bodies, and politicians.[5] He has little to say, however, about the relationship between conservation and field sports. Similarly, Emma Griffin's history of field sports in Britain, while it offers a compelling account of the constant adaptation of hunting to social and cultural changes, proceeds with little acknowledgement of the historical ties between sport and conservation or the extensive dialogues between these nature-cultures.[6] Where hunting, field sports, and conservation have been studied together, the focus has tended to be on game animals, particularly in the context of hunting and conservation in southern, eastern, and central Africa. In both John Mackenzie's study of European (especially British) hunting in the age of Empire and William Adams's history of conservation, it was the desire of British imperial hunters to protect their sport and its game species from local, African hunters that drove the creation of wildlife reserves and sanctuaries.[7] These protected areas were central to the maintenance of sporting culture, and they bequeathed a model of preservation and conservation that influenced wildlife conservation more broadly through the twentieth century.[8]

In a different setting, Louis S. Warren has shown how white recreational hunters in the United States and Canada were highly effective at lobbying for the passing of game laws and the creation of state-owned parks and reserves that marginalized and criminalized not only native American hunting, but also market hunting and other local forms of subsistence hunting. In this way, sportsmen conservationists helped to shape new ideas of 'national commons' that undercut other, older local uses of the land and its natural resources.[9]

While the work of Warren, Adams, and Mackenzie shows the value of studying the historical relations between hunting and conservation, all three authors pitch their analyses at the level of competing interest groups and have much less to say about the subjective attachments to wildlife found within the different nature-cultures they discuss. In thinking dialogically about the relations between birdwatching, field sports, and food customs, the account that follows seeks to open up the subjective and emotional dimensions of bird-centred practices and to reveal their often-similar ordering of the human passions in relation to wild birds. Thus, as we will see, falconers like the author T.H. White spoke of their love of their birds and how these relationships allowed them to play out a number of emotion-laden roles – from mother to teacher – in the training of hawks. Similarly, wildfowlers like the young Peter Scott not only thrilled to the sights and sounds of wild geese, but also formed maternal relations with some of his geese: one, whom he called Annabel, he saw as his 'goose child'. In a different way, bird harvesters in the Mediterranean confessed to the passion they felt for birds, and of being overwhelmed by their desire to catch and kill migrating songbirds and other species. In not dissimilar terms, birdwatcher Roger Tory Peterson revealed how, as a teenager, the slightest glimpse of a bird would provoke in him a 'fierce intensity'. Human–avian relations across the different bird-centred practices were brought to life by deep human feelings for them. A reflection on how these emotions were central to the attachments made to wild birds is at the heart of my account.

Some of the passions for birds found within birdwatching, field sports, and bird harvesting were bound up with the relationship between birds and the landscapes and environments in which people encountered them. In exploring the ordering of human–avian relations, the book reflects on the spatial dimensions and the different spatial scales at which these relations were enacted. This includes attending to the role played by wild birds in shaping human relations with local, national, and transnational landscapes and environments. Such a move draws on work within animal geography that has foregrounded the 'beastly spaces' and places of human–animal encounters, including 'nature regions', national landscapes, and across transnational spaces.[10] In what follows, I explore the links between wild birds and a range of landscapes, drawing on examples from the geography of 'wild' and suburban America, nature regions within the British Isles, and across the North Atlantic islands and Mediterranean. This attention to the ordering of human–avian relations across these different spaces of encountering and imagining wild birds challenges the near-obsession within twentieth-century natural history and field sports writing with placing birds within the geography of the nation and of seeing them, as writers like Julian Huxley did, as expressions of a national nature. Wild birds, however, lived and continue to live mobile, cosmopolitan lives that exceed and transgress the associations they have for particular human communities with the local geography and seasons of their homelands.

This sense of the cosmopolitan lives of birds and the international flyways they move along has not only been important to animal geographers and ethno-ornithologists, but, in the twentieth century, it has increasingly pushed sportspeople, birdwatchers, and conservation organizations to think about birds beyond the borders of nation states and national nature. If my book is informed by this recognition of the mobile lives of birds, it also seeks to place birdwatching, field sports, and food customs within an international frame. Thus, while the account developed centres on the British experience, this is a Britain stretched to include the relations between birdwatchers, conservationists, sportsmen, and hunters across Western Europe and the wider North Atlantic world. Pioneers of the new birdwatching like Huxley, Fisher, and Tory Peterson travelled widely across these regions, as well as further afield. More importantly, the ideas they drew upon and helped to shape also travelled across the North Atlantic world, helping to build international knowledge and shared practices. Ways of watching birds, as well as ways of protecting them, were shared across the Atlantic and the wider world, mirroring the mobile lives of birds. Similarly, wildfowling and falconry operated across international networks, linking participants in Britain with those in America, Europe, and (in the case of falconry) North Africa and the Middle East.

In thinking about bird-centred cultures across the different scales of local, national, and transnational space, the book inevitably draws on the work of geographers. My account also takes from this discipline an attention to the moral geography of outdoor recreations and the shaping of human conduct within different versions of the landscape, whether these spaces are conceived primarily in sporting, harvesting, or natural history terms. David Matless's work on landscape and associated human–animal encounters has been a major point of reference in this regard. From *Landscape and Englishness* to *In the Nature of Landscape*, and across many essays, Matless has detailed the 'contested pleasures' of different cultures of landscape associated with conservation, popular recreations, and sporting interests.[11] His attention to the moral geographies of leisure makes explicit what has often been implicit in accounts of nature conservation, natural history, and field sports: namely, the models of human conduct produced alongside the non-human.[12] My book documents this shaping of human conduct within diverse bird-centred practices, exploring the ways in which being human was styled in distinctive ways in relation to the avian within birdwatching, field sports, and bird harvesting. This includes detailing the shaping of spectatorship on bird reserves; the comportments and dress codes of naturalists, bird hunters, wildfowlers, and falconers; and the techniques and skills deployed by these various groups. These different ways of being human were not totally segregated from each other, and they were played out within 'avian landscapes' cohabited for much of the twentieth century by sportspeople, birdwatchers, and bird harvesters.

In seeking to capture the distinctiveness and the similarities of the styling and ordering of human–avian relations within these bird-centred practices, the book aims to understand the forms of attachment and entanglement between people and wild birds across these practices. While these attachments were emotional and subjective, and involved the shaping of human (embodied) conduct, they were also dependent upon a set of material and technological devices. Distinctive technologies brought birds closer to people, be they cameras, binoculars, hides, guns, nets, or ropes. These devices and technologies constituted the material dimensions of the 'contact zones' between people and birds through which, as Haraway puts it, 'species meet'.[13] They formed the material entanglements through which people entered into the 'more than human worlds' with wild birds. My book details how designed objects, built forms, texts, technologies, and manipulated landscapes worked together to form an assemblage or set of devices that shaped human–avian relations.

In exploring these emotional, technical, and material forms of attachment between people and wild birds within birdwatching, field sports, and bird harvesting, the book focuses on what I call the short twentieth century, beginning in the early decades of the century and running through to the late 1970s. In exploring change – as well as some of the continuities – in these years, the book does not follow a strict chronological progression but instead moves across this period in each of its chapters. These were years in which a new kind of natural history was forged and when new ideas of conservation – including the role of the state in the preservation of nature – gained a greater force and currency. It was also a period marked by the strengthening of conservation law in the United Kingdom, United States, and Western Europe, and by the growing influence of conservation groups within the policymaking process at the national and international level.

The strengthening of conservation law and policy owed much to the success of naturalists, conservationists, and field sports groups in making visible the threats to bird populations. These threats were certainly significant. From the 1920s to the late 1970s, a distinctive concatenation of forces drove profound environmental change in Britain, Western Europe, and North America. This included the drainage and reclamation of land as part of the wider intensification of agriculture; the catastrophic effects of agrichemicals upon ecosystems; the increased development pressure from industry, housing, tourism, and transport; and the impact of oil pollution on inshore waters and the coast.

If the rising tide of legal protection was a response to the growing awareness of this multifaceted environmental crisis, bird protection and conservation organizations themselves benefited from public concerns about the decline of bird populations and wider environmental change. In Britain, the Royal Society for the Protection of Birds (RSPB), the country's largest bird conservation charity, saw its membership soar from the end of the 1950s to pass half a million members by the early 1980s (reaching 1 million by the late 1990s). Internationally, bird conservation organizations also flourished, with the National Association of Audubon Societies (NAAS) in the United States enjoying substantial growth in its membership between the 1950s and 1980s. In Western Europe, birdwatching cultures expanded in Scandinavia, the Netherlands, and West Germany during this period. The growth of an international community of bird conservationists helped to strengthen the influence of organizations like the RSPB, the International Council for Bird Preservation (ICBP), and the World Wildlife Fund (WWF) in their dealings not only with national governments but also with supranational entities such as the European Economic Community (EEC) and the European Union (EU). Bird conservationists increasingly worked across national borders to protect the geographically mobile lives of birds and their international flyways.[14]

It was not only birdwatching and bird conservation that flourished through the short twentieth century: field sports also enjoyed a resurgence. The membership of organizations such as the British Falconry Club (BFC), the Falconer's Association of North America, the Wildfowlers Association of Great Britain and Ireland (WAGBI), and Ducks Unlimited (the North American wildfowlers body) grew strongly in the 1950s, 1960s, and 1970s, drawing in a wider swathe of both rural and urban society as supporters and followers.[15] Like the bird conservation and natural history societies, field sports benefited from growing affluence, mobility, and leisure time in the post-war decades, which gave more people the time and money to pursue these outdoor recreations.

While the diverse bird-centred nature-cultures all prospered during the post-war decades, the relationships between them were not only dialogical and disputatious but also often socially close. Cross-cutting social networks shaped the recreational fields of sporting and conservationist organizations from the 1920s to the late 1970s. In Britain, these were networks dominated by and under the leadership of aristocratic and upper-middle-class men. Individuals from this class cultural formation led the principal organizations and constitute the recurring caste of characters in this book. They include bird conservationist figures such as Max Nicholson and Peter Scott, whose upper-middle-class backgrounds were shared by leading figures within British wildfowling and falconry such as James Wentworth Day, Jack Mavrogordato, and T.H. White.

Historians and sociologists of post-war Britain have emphasized the relative decline of upper-class authority and the challenges to established social elites from the reorganization of social hierarchies from the 1940s to the 1970s. These were changes wrought by, among other things, the reforms of the post-war Labour government and by growing popular prosperity. Yet in the spheres of conservation, natural history, and field sports, those from elite backgrounds continued to exert considerable influence and leadership. As in some other areas of social life, including the world of contemporary media and entertainment, upper-class roles and identities were recast in forward-looking ways.[16]

In natural history and bird conservation, many elite men, like Huxley and Fisher, were driven by a desire to shape a new self-consciously modern form of ornithology and by an enthusiasm for the principles of planning and state-sponsored conservation. Many were committed 'moderns', who jostled with the anti-modern impulses of similarly elite men such as James Wentworth Day and others central to the promotion of field sports. They all sought to use modern means of communication – radio, photography, film, and mass-circulation print culture – to promote their ideas and, particularly among the new naturalists and bird protectionists, to reach a broader lay public beyond the landed elite and the upper middle classes. For Fisher and Huxley this was a genuinely cross-class public involving 'ordinary people' committed to the value of science, technical knowledge, and self-improvement.

The democratization of birdwatching and, to a lesser extent, field sports by the early 1970s, however, put pressure on the elite leadership of the various bird-centred organizations. One consequence of this process of democratization was to bring more urban and suburban enthusiasts into these forms of recreation. In the case of organizations like the RSPB this began to unravel their accommodation with sports like wildfowling and the privileged access to, and knowledge of, the countryside that had cemented the historical ties between those at the upper echelons of conservation and sport. Growing ethical differences fuelled by cultural distance generated sharper divisions between these bird-centred practices. This marked the unravelling of the close, constitutive relations between the nature-cultures of field sports and conservation, in particular, that had been a feature of their relationship during the short twentieth century.

In capturing the multiple and dialogical ways in which human–avian relations were organized and lived, I have drawn upon a diverse range of sources and forms of evidence. These include parliamentary records and newspaper reports; policy statements and internal communications from leading staff across sporting and conservation organizations; lectures, talks, and correspondence; together with autobiographies, biographies, and literary nature writing. These latter forms, especially nature writing, have expanded the canvas of the book, allowing more elaborated and highly wrought cultural reflections to be read next to prosaic written sources and archival evidence. All these diverse sources reveal attitudes and feelings towards wild birds and the wider natural environment.

The first four chapters of the book are concerned with the rise of the new bird science, bird study, and birdwatching. In chapter 1, I explore the formation and growth of the new ornithology in the first half of the twentieth century, tracing its institutional development via the creation of the British Trust for Ornithology (BTO) and the International Council for Bird Preservation (ICBP), and the innovations in US bird science. The chapter details the observational and recording practices associated with the new ornithology and its wider influence on popular natural history. In doing so, the chapter seeks to capture the ways in which elements of the old bird science survived in reconfigured forms within the new ornithology. This included a shared attention to breeding biology, the reworking of a collecting ethos, and the continued reliance on the use of museum studies of skins and eggs in the practising of the new science drawn from the taxonomic studies of the nineteenth century. For all its rationalist ambitions and attention to the collecting of data, the new ornithology, like the old bird science, mixed science and sentiment in its shaping of human–avian relations.

Chapter 2 charts the rise of birdwatching as a national pastime in Britain, focusing on the role played by the RSPB in popularizing the principles of the new ornithology and, through this, organizing the feelings for birds into a distinctive recreational culture. The chapter rejoins the argument made in chapter 1, seeking to explore the different kinds of attachment to wild birds facilitated by distinct bird-centred cultures of nature and the dialogue between them. In chapter 2, I contrast the new post-war birdwatching with the ongoing activity of bird's-nesting and the collecting of wild birds' eggs. While the RSPB sought to steer this impulse to collect into the co-operative practices of bird study, the chapter shows how the society had not only to dismantle the culture of egg collecting, but also to negotiate the continuing importance among the new birdwatchers of the post-war years of the sport of spotting and the collecting of 'lists' of birds. These were forms of attention to birds that echoed the preoccupations central to the old nature-cultures.

My exploration of the ways in which bird conservation organizations shaped popular attachments to wild birds continues in chapter 3, which details the way the RSPB sought to shape close-up encounters with wild birds on its nature reserves. Through the shaping of the landscape – including the erection of hides, screens, and signage, and the building of footpaths – the RSPB facilitated new ways of looking at wild birds and coming into contact with them. The chapter focuses on the RSPB's Minsmere reserve in Suffolk, where many of the society's ideas about bringing people and birds together were pioneered, and it considers how the society's attempts to shape ways of looking at birds were informed by ethological ideas about bird behaviour as well as being cut across by other ways of ordering human–avian relations.

If organizations such as the RSPB aimed to shape new pleasures in looking at wild birds in the post-war years, the boom in birdwatching during this period was increasingly shadowed by evidence of new threats to wild bird populations. In chapter 4 I explore how bird conservation organizations and conservation-minded scientists made visible the new threats to birdlife from post-war economic development and linked these threats to an understanding of a wider crisis of the natural environment within Britain and North America from the late 1950s on. Rachel Carson's influential book *Silent Spring* was at the heart of this campaign. The chapter sets out Carson's key evidence about the declines in wild bird populations in both Britain and America, and it argues that her work owed much to the efforts of amateur birdwatchers and conservation groups as well as to other professional biologists. Carson's writing, together with that of her interlocutors and the ordinary birdwatchers who wrote to her, painted a powerful, emotional picture of the disappearance of familiar birds. In both 'wild America' and its suburban backyards, and in the British countryside, the environmental crisis worked to reconfigure human–avian relations. While wild birds continued to evoke great pleasure, joy, and a sense of wonder, this was now mingled with sadness, melancholia, and feelings of loss. Among the RSPB's members in the United Kingdom and the Audubonites in the United States, looking at birds became entangled with the shadow of death, by population decline, by distressing encounters with mass bird casualties, and by landscapes transformed by the loss of birds.

The book's last three chapters shift from the world of bird study, birdwatching, and environmental crisis to the relationship between conservation-minded birdwatching and other culturally significant ways in which people sought out encounters with wild birds through the central decades of the twentieth century – encounters staged within field sports and the residual traditions of harvesting wild birds for food. Chapters 5 and 6 focus on, respectively, the field sports of wildfowling and falconry. Chapter 7 looks at bird harvesting in Britain, the North Atlantic world, and Northern and Southern Europe. In exploring these bloodier relations with wild birds, the last three chapters of the book show how each of these practices was in a sustained dialogue with bird conservation and the new birdwatching. Out of this dialogue came not only changes to the law and a move by field sports and bird harvesters to align their practices with bird conservation, but also, as I show, evidence of common passions for birds and similarities in the styling of the human in relation to the avian, including the shared material devices for bringer wild birds closer to people. Central to all three chapters is an attention to the aesthetic, emotional, and technological attachments to wild birds within these bird-centred practices. Across these practices, acts of love and care were combined with the visceral pleasures of hunting and killing birds. All of these ways of encountering wild birds allowed an enlargement of the human experience, with the falconer, wildfowler, and bird hunter inhabiting more-than-human worlds through their intense encounters with wild birds.

1

'A Creed for Bird Watching': Bird Study and the New Bird Science

Some of the fundamental problems of bird life resolve themselves, upon analysis, into simple questions of numbers.

E.M. Nicholson, *The Study of Birds* (1929, 56)

In 1931, Tom Harrisson, a twenty-year-old undergraduate at Pembroke College, Cambridge, and Philip Hollom, Harrisson's junior by a year, organized a national enquiry into the great crested grebe, a common species of water bird. The grebe had experienced a dramatic decline in its fortunes during the second half of the nineteenth century and had nearly been driven to extinction by the demand for its feathers in the millinery trade. By the first decades of the twentieth century, however, it had undergone a significant revival.

Harrisson and Hollom's inquiry set out to document and explain the grebe's resurgence. Rather than simply noting the bird's movement from being rare to relatively common, they sought to precisely measure the number of breeding pairs in England and Wales. Following the model of the smaller census of heronries organized by the journal *British Birds* in 1928, in which Harrisson had participated as a schoolboy, the two young enthusiasts mobilized 1,300 volunteers to establish the distribution of the bird during the spring and summer of 1931.[1] The volunteers were recruited by the placing of letters in the national, local, and sporting press, with the BBC also broadcasting an appeal at the beginning of its general news service.[2] The inquiry was organized on a county basis, with observers required to identify and visit suitable bodies of water during the breeding season. They submitted their results in tabular form, giving the precise location of the water, the number of breeding pairs, and the year in which the water was colonized.

The Times celebrated the success of the inquiry in a review of the final report when it was published in October 1932, congratulating all the observers who took part in 'this national effort of field ornithology'.[3] The success of the census bolstered the thinking of a small group of ornithologists and intellectuals, many of whom had participated in the grebe inquiry, who had already begun to fashion a new popular science of bird study. They defined this against the established scientific ornithology, seeking to place the living bird and its life history at the heart of a 'new ornithology'.

This chapter sets out to explore the formation and growth of this new form of bird study and birdwatching, tracing its institutional development via the formation of the British Trust for Ornithology (BTO), the innovations in US bird science, and the development of the International Council for Bird Preservation (ICBP). The chapter explores the novel observational and recording practices of the new ornithology and sets out its wider influence on popular natural history.[4] In doing so, my account draws upon and seeks to revise established histories of ornithology and popular natural history.[5] These histories have emphasized how the new bird science and the 'new birdwatcher' were distanced from both sentimental bird lovers and the older tradition of bird collecting and museum study. Authors like Matless, Toogood, and Macdonald have closely followed the modernizing narrative of the 'new ornithologists', ventriloquizing their self-positioning and self-justification. Because of this they underplay the common preoccupations that linked the new ornithology with its precursor.[6]

In the account that follows, I develop a different argument about historical change within bird science – one that emphasizes the lack of neat breaks and that seeks to capture the way in which elements of the old bird science survived in reconfigured forms within the new ornithology. This occurred in four distinct ways. First, the new ornithology shared with the old a focus upon birds' nests and breeding biology. One consequence of this was that the skills of nest finding that are so central to egg collecting persisted in the study of breeding biology within the new bird study. Second, the practices of collecting and organized pursuit (or 'hunting') of birds within the 'old ornithology' were redirected into the collecting of data about birds and their life histories within the new ornithology. Third, for all its criticisms of collecting, the new bird science continued to make use of museum studies of skins and eggs in the practising of the new science and was indebted to the taxonomic studies of the nineteenth century. Fourth, both traditions of bird science mixed the empirical and the sentimental. An aesthetic and emotional connection with wild birds and their eggs certainly marked the older bird science. As one observer has noted, among the bird collectors there was a deep 'passion for beauty and the lust for curiosities'.[7] Similar feelings to these, as well as other kinds of emotional investment and sentimental connection with birds, entangled the rationalist and scientific ambitions of the new ornithology. Science and sentiment were not easy to separate. Wild birds, and the environments they inhabited, continued to stir the emotions of the 'new ornithologists'. Even for the most rationalist quantifier of wild birds, engaging with them was about more than simply counting them. In all these ways, then, the older practices and techniques of collecting continued to inform – to haunt – the new popular science of bird study.

Alongside exploring how the new ornithology developed by reconfiguring elements of the old bird science and its preoccupations as much as inventing new methods and approaches, the chapter also seeks to bring a more international dimension to the account of the new ornithology. Histories of ornithology like those developed by Matless, Macdonald, Toogood, Bircham, and Evans have shared with their object a focus on the nation as the predominant scale of both organization and meaning.[8] The new ornithology forged in Britain in the first half of the twentieth century, however, was not unique. It was shaped by the flow of ideas between Britain, Europe, and North America. As Mark V. Barrow has shown, since the 1850s, and especially in the post-Civil War period, American ornithologists consistently looked towards developments in Europe in establishing the scientific credentials of ornithology. This had initially involved the adoption of taxonomic and faunistics studies, but from the end of the nineteenth century and into the early years of the twentieth it included a common concern with forging networks of birdwatchers collecting data on distribution, migration, and life history.[9] The British proponents of the new ornithology had their work reviewed and also published in journals like *The Auk*, produced by the American Ornithologists Union (AOU). They also drew inspiration from the pioneering national census of birds conducted by the Bureau of Biological Survey in the United States.[10] At the same time, British ornithologists, like their American counterparts, looked towards developments in bird science on the continent of Europe, especially in Germany and Holland, where pioneering work on bird migration and bird ringing had been undertaken. In this sense, ideas about bird study, like those about bird protection, moved across the North Atlantic world from the last years of the nineteenth century and especially in the first half of the twentieth. They were carried through not just ornithological journals but also through the formation of bodies like the International Committee for Bird Preservation (ICBP) (established in 1922). The ICBP was a body formed of Atlantic crossings, established at a meeting in London but initiated by a leading figure in the American National Association of Audubon Societies and involving continental European ornithologists. The founding of the ICBP helped not only to internationalize bird protection, but to establish the new ornithology as a genuinely international practice for bird observation and study, though one that was tellingly focused in each country principally around national avifauna.

In the first section of the chapter I explore the formation of the new ornithology, setting out the ideas that shaped it and the role played in particular by the British Trust for Ornithology (BTO) in providing an institutional home for this self-consciously modern vision of birdwatching. This section places the conception of bird study promoted by the BTO next to the established traditions of 'scientific ornithology', and it reflects on the conflict between the supporters of taxonomy and faunistics and the 'new science'. Egg collecting became a particularly contentious subject – one that divided ornithologists – in the early decades of the twentieth century. Bird protection organizations like the Royal Society for the Protection of Birds (RSPB) moved towards a more assertive advocacy of stronger bird protection laws, including protection of birds' eggs, in the 1920s and 1930s. Through this they sought to draw stronger lines between legitimate birdwatching and the dubious practices of the collectors. This exposed tensions between the cause of bird protection and the new bird science as much as the old. Up until the 1940s, divisions persisted between the RSPB's protectionist approach, and its support for the 'bird lover', and the BTO's embrace of the new science, including the contested practices of bird ringing and nest counting.

In the chapter's second section I consider the role of publishers in promoting the new style of natural history and bird study to a wider audience. I focus on a number of early influential bird studies produced within the Collins New Naturalist series, together with writing published outside the series. This section explores the work of three authors in particular: John Buxton, James Fisher, and Desmond Nethersole Thompson. All three were exemplary 'new naturalists' concerned with wild birds and seeking to elaborate a new style both of doing bird study and of writing popular natural history. All three authors mixed the new bird science with an aesthetic, landscape sensibility. They did this even as they defended the new science against its critics, who worried that the new ornithology was a soulless, emotionally cold pursuit. The 'new naturalists', and especially Nethersole Thompson, also remained caught up in and committed to elements of the older traditions of the bird collectors.

A NEW SCIENCE OF BIRDWATCHING

In 1932 Max Nicholson published *The Art of Bird Watching*, subtitled 'A Practical Guide to Field Observation'. It was the twenty-nine-year-old Nicholson's third book, and it was a statement of intent. Nicholson was a young journalist at the time, but he was emerging as a key protagonist in the shaping of the idea of state-sponsored conservation and of a 'planned countryside'.[11] He was associated with the influential think tank Political and Economic Planning (PEP) (formed in 1931) and would later play a key role in post-war conservation as the director of Nature Conservancy (1952–66) and as president of the RSPB (1980–85).

The Art of Bird Watching set out a manifesto for a new kind of birdwatcher. At its core was a plea for a more organized and scientific approach. No longer should the birdwatcher be someone 'who rambles about the countryside until chance puts something in his way'. Rather they should develop an 'intense concentration of scope'.[12] This might mean focusing on a single bird community or attention towards specific areas of behaviour like courtship or migration. At the heart of this approach were news styles of observation. Much of *The Art of Bird Watching* was dedicated to setting out Nicholson's delineation of a carefully coded set of observational practices. The book offered a vision of how new technologies and devices could transform birdwatching. These included the use of telescopes and binoculars – especially the new lighter tourist field glasses – for field observation, as well as the use of hides and observation posts. These devices brought the bird closer to the observer without the need to shoot them for identification or collecting purposes. Nicholson even floated the possibility of using motorcars as movable hides and of deploying gliders and aircraft to study the movements of soaring birds.[13]

Other modernizers within the world of ornithology shared his enthusiasm for new technologies. The bird collector, soldier, and spy Colonel Richard Meinertzhagen, responding to a proposal from Nicholson to study the flight speeds of birds, suggested using cars, motorcycles, and aeroplanes in order to estimate their velocity. Tom Harrisson, co-author of the great crested grebe inquiry, confessed that he 'also checked a lot of bird speeds again, from some peculiar vehicles!' while on a trip to Ireland.[14]

While other birdwatching moderns shared Nicholson's enthusiasm for new technology, there was one piece of more humdrum equipment that was key to the art of the new birdwatching: the notebook, and the associated art of note taking. While conceding that 'any kind of notebook will do for bird watching', Nicholson argued that it was imperative to record observations of bird numbers and behaviour in a notebook as soon as they were seen, with the full field record noting times and weather conditions. Diagrams, sketches, and rough maps should also be used, and field notes should later be written up and placed in a loose-leaf book or a card index to create a more permanent record.[15]

To be of greater value, these observations should also form part of a larger collective effort. Bird census work was a key vehicle for this coordination of the records of single observers. It was central to the scientific ambitions of the new ornithology promoted by Nicholson. Census work – such as that pioneered by the journal *British Birds* and by Harrisson and Hollom – usually counted breeding birds within a specific geographical area. A map was essential for every census and Nicholson recommended the Ordnance Survey six-inch to the mile maps of England and Wales.[16] The census area was produced by dividing up a given geographical location into a grid of equal squares and then counting the birds – breeding pairs, nest, or singing males – within this square. Chapter 4 of *The Art of Bird Watching* offered a manifesto for how these close observations of birds could be moulded into a 'National Society of Birdwatchers'.[17] As Nicholson argued, 'whether he liked it or not, the birdwatcher is engaged in a cooperative undertaking… There is no such thing as an independent bird watcher ... there are only organized ... or disorganized ones'.[18]

The Art of Bird Watching was enthusiastically reviewed in *The Auk*, the journal of the American Ornithologists Union. For the reviewer, Nicholson's book would assist in the development of 'well-trained bird-watchers, field students who will know what to look for and how to record their research'. It offered a guide to 'a valuable method of scientific research' and was almost the first to 'point the way and lay down the rules of the game'.[19] *The Auk's* positive reception for Nicholson's book was evidence of the convergence of thinking from those interested in forging a new ornithology on both sides of the Atlantic. Developments in bird migration study, bird banding (ringing), and bird counts had occurred in the United States from the late nineteenth century, with the Biological Survey playing an important role in these studies.[20] In 1900, Frank Chapman, publisher of the US journal *Bird-lore* (1899) and of many popular bird books in the early years of the twentieth century, proposed the first Christmas Bird Census to replace the traditional Christmas Day hunt. By the mid-1930s, the census had become a significant tradition within American birdwatching. It was complemented from 1937 by an annual spring count of breeding birds organized by the National Association of Audubon Societies.[21]

The question of field identification and 'sight records' also became a central preoccupation for American ornithologists. As more birdwatchers went into the field to count and record birds, demand grew for reliable field guides that provided the means to identify birds in the field rather than once they were dead. In 1923, Ludlow Griscom produced the first field guide of American birds that focused on those found in New York county, guiding his readers towards field marks rather than offering detailed descriptions of specimens in the hand.[22]

These different elements began to cohere into a version of new ornithology that shared many features with that proposed by Nicholson. Bird study was also central to the vision of international ornithology promoted by the International Committee on Bird Preservation (ICBP). Formed in London in 1922, the ICBP owed much to the efforts of leading figures in US ornithology and sought to 'cultivate through the World ... appreciation of the value of the living bird'. This worldwide cooperation was underpinned by the exchange of literature promoting not just bird protection, but 'bird study'.[23]

In the United Kingdom, Nicholson's vision of the new ornithology was developed and given institutional direction by the organization that he helped to found in 1933: the British Trust for Ornithology (BTO). Nicholson and others – such as Julian Huxley, the eminent zoologist and birdwatcher – conceived of the BTO as an 'independent clearing house and directive centre for all kinds of ornithological fieldwork in the United Kingdom'. A national council would help to coordinate 'a chain of organized bird watchers throughout the British isles, with correspondences in the Empire and abroad'.[24] As both Helen Macdonald and Mark Toogood have argued, the recruitment and coordination of amateur observers by the BTO shared much with other networks of amateur observation in the inter-war and wartime years, most notably Mass Observation (MO).[25]

There were even strong personal and practical connections between the two forms of 'mass observing'. Not only was Tom Harrisson, the key founder of MO, a keen birdwatcher, but he had also, as we have seen, co-organized the first national census of great crested grebes.[26] The latter study not only prefigured the inquiries of the BTO but, in mobilizing amateur observers and with its faith in the social value of modern science, it also became a model for Harrisson's approach to mass observation and his interest in the behaviour and social rituals of the British population. The two forms of mass observation remained close in Harrisson's thinking. In a wartime letter to James Fisher, secretary of the BTO, Harrisson revealed that he was planning to use one of Fisher's bird studies as the basis for an article in the *Forces Paper* in the Mediterranean. It would, he told Fisher, link up his study 'with the general application of counting and statistics to animals affairs, including man'.[27]

Other contemporary supporters of mass observation and the new ornithology saw the close intellectual and practical ties between the two movements. In his forward to Madge and Harrisson's manifesto for MO, Julian Huxley had noted this connection between the new birdwatching and mass observing. As he pointed out, both practices offered 'room for the untrained "amateur" just as readily as the trained scientist'.[28] In the same publication, Madge and Harrisson identified the question 'Why do you watch birds?' as one of a number of questions – including the intriguing 'What is on your mantelpiece?' and 'What do you mean by Freedom?' – that would be dealt with in a series of projected pamphlets.[29] From within the BTO the parallels were also clear. An internal memorandum from 1941 reflecting on a survey of starlings described the inquiry as 'an attempt at mass observation in a routine manner of certain habits of a very common bird'.[30] The ties between the new ornithology and MO were also cemented at an organizational level within MO itself. Both Max Nicholson and James Fisher were involved in the management of MO, alongside a colleague of Nicholson's from PEP, Richard Fitter, author of one of the early New Naturalist books on *The Birds of London*.[31]

If the BTO and many of the 'new ornithologists' were intellectually close to the form of democratic science promoted by mass observation, Nicholson's vision for the new ornithology was deliberately set against the tradition of museum study and the collection and display of bird skins that dominated established 'scientific ornithology' in the 1920s. One of Nicholson's key targets in this shaping of a new ornithology was the British Ornithologists Union (BOU). The BOU had been formed by a group of wealthy collectors in 1858, and it published the influential journal *The Ibis*. The Union focused on the description of bird taxonomy and faunistics (studies of geographical distribution) primarily of Indian and African birds.[32] As one of its critic observed in 1959, *The Ibis* was made up almost exclusively of papers on 'geographical ornithology' derived from private or military expeditions and dominated by long lists of collected specimens. Many of the authors were, as another commentator observed, typically colonels, rear admirals, or colonial office administrators.[33]

Nicholson himself was clear about the urgent need to break with this approach to wild birds. Writing to a colleague at Oxford in December 1933, he suggested that the BTO would 'bring together all who are interested in the bird while it is alive and free. Those who favour dead specimens [and] blown eggshells … are already liberally provided for.'[34] If forging a new kind of ornithology was important to Nicholson, he was careful not to alienate all the supporters of museum study and taxonomy. At a meeting of the Oxford Bird Census in May 1931, Nicholson suggested that his proposal for a 'new independent "National Ornithological Centre" was designed to be complementary … to the British Museum in the systematics sphere'.[35] One of the early meetings that led to the formation of the BTO was held in the boardroom of the Natural History Museum, the principal depository of bird skins and eggs in Britain, and was chaired by Dr P.R. Lowe, the head of its Bird Room.

Lowe himself was one of the eleven prominent signatories of a letter to *The Times* announcing the formation of the BTO and calling for subscribers and donors.[36] Nicholson's conciliatory attitude towards organizations like the Natural History Museum was part of the alliance building and practical politics that he knew was necessary to give the BTO a good start in life. As he confessed to Andrew Dawson, the editor of *The Times*, in explaining why it had taken some time to finalize all the names on the BTO appeal letter, 'ornithology is so riddled with its own politics that it is essential to have a thoroughly representative list, so we have them all in – Scots and English, oologists and protectionists, museum people and field men'.[37] As the BTO developed, however, it did not completely break with museum study and the collection of skins. It ran a study material service, for instance, and in 1949 it reminded its supporters that skins could be loaned for scientific or educational purposes. It also encouraged its members to send any freshly dead bird they found to the Trust as this would help with 'comparative studies and for replacements'.[38] In June 1932, the journal *British Birds*, which was close to the BTO and was later edited by Nicholson, regretted the sale of part of Lord Rothschild's collection of bird skins to the United States and encouraged its readers to help museum authorities fill the gaps in their collections if they went abroad or lived overseas.[39]

If the supporters of the new ornithology defended museum study, egg collecting was subject to more disapprobation, though here too the position of the promoters of the new science was complicated. Nicholson had railed against the practice of egg collecting in his second book, *Birds in England*, published in 1926. Confessing that he, like many of his contemporaries, had collected himself, he went on to explain why he had changed his mind and why he felt that the guilt of the collector 'has been and still is immense'. This was despite the 'ill-advised' and 'grossly exaggerated' 'protectionist propaganda' against it.[40] The latter had laid too much of the blame for the declines in bird populations on egg collectors, exaggerating their effect on the drop in numbers of both common and scarcer species. It was undoubtedly the case, however, Nicholson conceded, that for birds whose numbers were already at a low ebb and whose rarity value added to the appeal of their eggs collectors had pushed certain species towards extinction. *Birds of England* detailed the often-inflated scale of many egg collections, including those of R.W. Chase, who hosted an exhibition in the grounds of his house in 1924 that included some of the 12,000 eggshells (and 1,135 stuffed and mounted specimens) that he possessed. Nicholson also challenged the gloss of science that many collectors sought to give to their hobby. As he suggested, this pretension to science was nothing more than 'a stalking horse': 'they have contributed little more to our knowledge than stamp collectors, who are at least harmless'.[41]

It was not just Nicholson who was critical of egg collecting. Critics within the BOU had begun to challenge the practice. Like Nicholson, they questioned its scientific value as well as the obsessive focus on geographical variation within eggs of the same species and the taking of whole clutches. These criticisms led to a split within the BOU in which oologists formed a separate club, the British Oological Association (BOA), in order to meet and discuss their collections. This generated a journal, *The Oologists Record*, launched in 1921.[42] Even more so than *The Ibis*, *The Oologists Record* was dominated by contributions from men who were members of the established professions and retired servicemen.[43] The articles it published reported on collecting trips, often overseas, and detailed the eggs of rare species and unusual variations in the clutches of common birds. Its contributors regularly sought to emphasize the continued scientific value of oology, mixing claims to science with the sport and adventure of nest finding.[44]

The debates over the ethics of egg collecting within the BOU and among advocates of the new ornithology echoed those that had already begun to surface in popular books around the turn of the century and had later spilled over into the broadsheet press. In the preface to their book *Our Rarer British Breeding Birds, Their Nests, Eggs and Summer Haunts*, the pioneering bird photographers Richard Kearton and Cherry Kearton worried about the threat posed by egg collectors to rare breeding birds, fearing that the legal protection for rare birds was too weak.[45] This was an important caveat coming from such popular authors, particularly as their earlier book *British Birds' Nests: How, Where, and When to Find and Identify Them* was effectively an egg collectors' manual.[46]

While the Kearton brothers' comments were certainly significant, more important still was the robust support given to the cause of bird protection by the leading broadsheet newspaper *The Times*. By the early 1920s the paper was editorially sympathetic to bird protection and provided a platform for oology's critics. A leader article from July 1923 suggested that 'public conscience' was increasingly sympathetic to the protectionists and hostile to those who killed rare birds and took their eggs. This 'humane opinion' was, in particular, strongly against those who 'masquerade as servants of science'. 'A good many crimes are committed,' it concluded, 'in the name of oology.'[47] Other critics of egg collecting agreed. Between 1936 and 1938 the letters pages of *The Times* were the setting for duelling exchanges between the critics of collecting and its defenders.[48] Under the heading 'British ornithology', Colonel (Rtrd) B.H. Ryves, founder of the Cornwall Bird Watching and Preservation Society, led the attack on collecting. He contrasted the scourge of collectors with the scientific benefits to the nation of the new ornithology. 'Is it not a scandal,' he suggested, 'that these despoilers … are allowed to … persecute the birds which are the nation's property to be enjoyed by all, and cause stagnation in ornithology.'[49]

Other vociferous critics, including Nat Tracy from Kings Lynn, supported Ryves's letter. Tracy underlined the 'evils of egg collecting' and revealed that he was 'working on a scheme' to put an end to it. True to his word, Tracy established a new organization and campaign: the Association of Bird Watchers and Wardens. *The Times* duly featured its establishment in a short, sympathetic news item.[50] The Association of Bird Watchers and Wardens would become a key opponent of egg collectors through the late 1930s and 1940s, developing strong ties with the RSPB, the main bird protection charity. Geoffrey Dent, the head of the RSPB's own 'Watchers Committee' in the 1940s, was a member of the association and engineered a 'nest adoption' scheme for birds of prey that was later taken up by the RSPB.[51]

Concerned to build alliances across ornithology, the BTO's official policy towards egg collecting in the 1930s was publically agnostic. Stung by criticisms of it in the press amid the extensive debate about the problem of egg collecting, the BTO's Council issued a tortured statement on the subject in its second annual report. This defended the need to remain above the politics of egg collecting so as not to jeopardize its role as a 'fact-finding service institution for all bird-watchers'.[52] It tartly noted that other organizations were happy to make pronouncements on the subject and critics should therefore direct their concerns to those bodies.[53]

The BTO's decision to condone egg collecting in order to preserve the cooperation of all ornithologists, including oologists, stemmed in part from the links some of its leading members had historically sustained with the collectors. Julian Huxley, for example – one of the co-founders of the BTO – had requested assistance on two occasions in the late 1920s from *The Oologists' Record*, asking for clutches of common birds' eggs to be sent to him for analysis following the harsh winter of early 1929.[54] The Trust's agnosticism, however, as well as aspects of the new bird science, caused tensions with protectionists.

While the RSPB had given support to the BTO at the time of its formation, many within the Society were sceptical about some of the new scientific methods promoted by the BTO and the potential risks they posed to the safety of birds. In particular, nest record schemes and bird ringing drew disapprobation from some of the RSPB's supporters and staff. A short article that appeared in *Birds Notes and News* (the RSPB's members' magazine) questioned the value of the growing activity of bird ringing or banding. A bird ringing scheme had been established by the journal *British Birds* soon after its establishment in 1907. As Max Nicholson later noted, the journal and its editor–publisher, W.F. Witherby, were important rallying points for those concerned with developing studies of living birds.[55] The *British Birds* ringing scheme drew inspiration from the pioneering ringing of birds by Heinrich Gatke on Heligoland, on the North Sea coast of Germany, from the 1890s. The locals of Heligoland had long caught birds on the island, and Gatke transformed this practice of catching into scientific study. He devised a large bird trap, superseding the islanders' simpler traps, with a funnel entrance that narrowed to a catching box from which the birds could be collected, ringed, and released. This was the Heligoland trap, which was first exported to Britain in the 1930s and became a mainstay of the British bird observatories after the Second World War and an iconic device of the new ornithology.

For the contributors to *Bird Notes and News*, however, bird ringing had limited value and carried risks to the welfare of birds. Seeing the practice as a symptom of the 'tabulating industry of the present age', one article conceded that ringing had some value as a tool in bird migration studies. However, its methods of trapping were likely to cause distress to the captured birds, with some even being injured or dying. When the pointless repeated catching, ringing, and releasing of sedentary species was added to the charge sheet, ringing had little to contribute to bird protection.[56]

The recording of nests – pioneered at Oxford University in the 1920s and later to become one of the BTO's longest-running inquiries – also provoked concern.[57] Julian Huxley, at the time a member of the RSPB council, was drawn into print on the pages of *Bird Notes and News* to defend a survey on the reproductive habits of blackbirds led by a colleague, Dr Baker, and supported by the BBC. Huxley described Baker's study as 'a deliberate attempt to see whether, in the field of open-air natural history, some kind of [innate curiosity of amateurs] could be … employed on a problem of genuine scientific interest'. The service of these amateurs was to be focused on collecting information on clutch sizes and nesting mortality. This corporate project would engage those who might otherwise 'confine themselves to an aimless collection of bird eggs or an equally aimless collection of records of rare species'.[58]

The editor of *Bird Notes and News* was not persuaded. While they endorsed the value of 'open air natural history', they worried that Dr Baker's study would 'stimulate the ever popular search for nests [and hence eggs] without setting forth [any useful evidence]'. The welfare of birds would be harmed, for no conceivable benefit.[59]

Other commentators shared this scepticism towards the new ornithology and its emphasis on the counting and quantification of birds. In the pages of the conservative country magazine *The Field*, writer and self-confessed 'countryman' Richard Perry challenged the work being done by *British Birds* and the BTO. For Perry, the new bird science was too rigid in its approach, denigrating those who emphasized any appreciation of the beauty of birds and their environments. As he put it, 'any writer who chooses to introduce any aesthetic feeling into his studies of birds is regarded with the greatest suspicion by the ornithological authorities [the BTO and *British Birds*]'.[60] He distanced himself from the new bird science, worrying that ornithology was 'being reduced to a cold presentation of statistics and trap records'. His appreciation of birds was, in contrast, that of a 'countryman's interest in the wildlife around him'. This grew out of living, year-in and year-out, with wild birds, knowing them intimately not as a '"biological corpse" but as a vital living entity'. This was the countryman's knowledge, and he feared that the BTO was turning the study of birds into a practice that was 'urbanized and bureaucratic'.[61]

Perry's article received support from others in the letters pages of *The Field*. Correspondents rallied to his idea of the centrality of aesthetic appreciation and sentiment in the observation of birds and the dangers of being 'overly-systematic' and turning birdwatching into a 'ruthless business proposition'.[62] Perry's own position, however, was closer to the new ornithology than he had cared to admit in his article. His book *At the Turn of the Tide: A Book of Wild Birds,* first published in 1938 and reprinted in 1943, fashioned a form of nature writing that was based on close field observation and recording.[63] Its evocative, highly personal account of encounters with wild geese, in particular, and the shifting landscapes of tide, estuary, and salt marsh was built up from rigorous fieldwork and observation. Perry's descriptions of birds and their behaviour owed more than a little to the 'art of bird watching' set out by Nicholson. As Perry noted in his book's introduction: 'I have aimed at a vivid portrayal from life of the birds of these wild places, but with absolute accuracy from my own observations.'[64] He also engaged with studies and theories of bird migration and accounts of bird emotion and behaviour.

Perry's indebtedness to the new ornithology was also evident in his second book, *Lundy: Isle of Puffins,* published in 1940. In a review for MO later that year, Tom Harrisson criticized the book for its failure to offer a more literary rendering of the observations upon which it was based. Harrisson saw the book as part of a genre of publishing in which field notes were transposed in an almost unmodified form, with negative consequences for the enjoyment of the book's readers.[65]

The possibility of holding together emotional reactions to birds and systematic observation was addressed by P.H. T Hartley, a member of the BTO, in another rejoinder to Perry's article. Hartley suggested that the scientific field ornithologist did feel emotions like joy in their encounters with wild birds. Emotion, however, had to be extracted from the systematic record to enable a clear focus on the bird being observed. As he put it: 'When he [the scientific birdwatcher] marks a singing skylark, his eye and ear are not less appreciative because he holds a stopwatch in his hand and will only enter the minutes and seconds in his notebook.'[66]

Hartley's insistence on the suppression of emotion and pleasure in the recording practices of the new ornithologists probably confirmed the views of the critics of the 'new science' that it missed what was most crucial in people's relationship to wild birds. As we will see in the next section, however, some of the best popular natural history writing about birds published between the late 1940s and the early 1970s, which drew strongly from the new ornithology, did affect a partial rapprochement between science and sentiment, detached observation, and subjective feeling.

NEW NATURALISTS AND THE STUDY OF BIRDS

In late 1945 the publisher Collins launched a new series of natural history books. Called the New Naturalist, the series sought to bring together developments in the observation and understanding of nature informed by the relatively new sciences of ethology and ecology.[67] Billy Collins, the editor and publisher at Collins, turned to Julian Huxley to help him realize this new kind of survey of Britain's natural history.

They met in 1942, along with Huxley's protégé James Fisher, to sketch out an outline for the series.[68] All agreed that the books should embrace not just the new observational sciences but also developments in photography, including colour photography, to record and show bird, insect, and plant life as it occurred in its 'natural habitat'. Leading bird photographer Eric Hosking, known personally to Fisher, was enlisted as the photographic editor for the series. His photographs would feature frequently in the New Naturalist bird monographs. Alongside the embrace of colour photography, the New Naturalist series further affirmed its modernity by the choice of striking lithographic cover designs. Produced by the graphic artists Clifford and Rosemary Ellis, they gave a strong visual identity to the new books and signaled the mission of the series to promote a new, self-consciously modern naturalism in Britain.[69]

The early New Naturalist books made an immediate impact. A later 'biographer' of the series described the years 1945 and 1946 as '*anni mirabili*' for the venture, with public appetite for the books exceeding Collins's expectations. The first four books sold out their print runs in the first year, and second editions saw each achieving sales of around 40,000 copies.[70] Buoyed by the success of the book series, a *New Naturalist* journal, edited by James Fisher, was launched in 1948. While the journal turned out to be short-lived, it helped to amplify the vision of 'new nature' set out in the books. Central to this was the ambition to 'serve field workers' and to be biased 'towards the living thing and its home in Britain and towards the study of evolution'.[71] Extending eighteenth-century naturalist Gilbert White's plea for the parochial study of natural history, Fisher saw the journal and the book series as the province of the 'Island biologists' who would publish their discoveries of 'these magnificent islands with their varied landscape and climate'. Overall, 'our ambition', as he memorably asserted, 'is to capture British natural history, and to transform it into lines of type and blocks and colored ink'.[72]

This desire to reinstil pride among the British in their national flora and fauna was echoed in the frontispiece to every New Naturalist book.[73] This environmental nationalism was undoubtedly coloured by the legacy of wartime anxieties and threats of invasion. It also spoke, as others have noted, to a broader process from the 1930s of turning inwards within British culture. This found expression in a wider rediscovery of Britain and its landscapes as the Imperial project and its geographical scope began to recede.[74] James Fisher had powerfully captured this wartime conception of Britain's island status in the preface to his best-selling Pelican book *Watching Birds*, first published in 1941. Reflecting on whether the British public really needed a book on birds in the midst of the nation's fight for survival, he had claimed that birds were 'part of the heritage we are fighting for'.[75] This attention to birds as an important part of the natural heritage of Britain was combined in Fisher's preface with a social democratic vision in which the 'new birdwatcher' took their place in Britain's 'New Jerusalem'. As he suggested, the post-war world would bring a more democratic culture of birdwatching into being, pushing aside the 'privileged few for whom ornithology is an indulgence'.[76] Fisher's preface to the 1946 printing of his book extended this appeal to a new ornithological public. They were those 'ordinary men and women' who were the new subscribers to the BTO and were interested in bird study – the same people who 'buy Penguins to read on trains'.[77] This link with the 'new birdwatcher' was underscored by the fact that *Watching Birds* came with a BTO membership form on its back cover.[78] Fisher's later Pelican book, *Bird Recognition*, also included the same BTO application form.[79]

If Fisher's *Watching Birds* confirmed his reputation as the 'Julian Huxley for the masses', condensing the observational principles of the new ornithology and the BTO's democratic science for a wide readership, Fisher's New Naturalist monograph on a seabird called the fulmar offered a larger canvas on which to set out the new way of observing and recording birds. It combined this, however, with an emotive and personal account of his relationship with a mysterious and curious bird.

WATCHING THE FULMAR

This book is the result of an obsession. I have been haunted by the fulmar for half my life; and have needed no spur to explore its history, and uncover its mysteries, save the ghost-grey bird itself, and green islands in grey seas... Since 1933 I have lived no summer season without a sight of at least some of the great cliffs, and fulmar colonies.

James Fisher, *The Fulmar* (1952, xiii)

James Fisher's New Naturalist monograph on the fulmar was a fat, comprehensive survey of the ecology, evolution, distribution, behaviour, lifecycle, predators, and food of the bird. It also commented on the Antarctic fulmar of the Southern oceans.[80] Much of the book was devoted to the remarkable extension of the range of the fulmar, particularly around the coast of Britain. From its sole breeding base on the isolated island of St Kilda in the first quarter of the nineteenth century, the bird had spread around Britain's east and west coasts. Fisher's knowledge of the spread took fifteen years to collect and was supported by 400 observers and correspondents to the BTO. Following the model of Harrisson and Hollom's great crested grebe census, the BTO questionnaire asked observers to record the location of what Fisher termed every 'fulmar station' they could find. George Waterston, later the head of the RSPB in Scotland, ran the survey from 1933, with Fisher joining him in 1934. This initial census was then extended in 1944 with further BTO support and assistance from the army and other service units.[81]

A map showing Britain and Ireland encircled by the spread of fulmars graphically captured the biological success of the birds in adapting to new feeding and nesting opportunities. Fisher also saw the map as a measure of the 'indefatigability and keenness of the British bird-watcher'. The number and social mix of the observers confirmed Fisher's belief that the new ornithology was a genuinely cross-class movement. Echoing his comments about the range of people whom he claimed liked to watch birds set out in chapter 1 of *Watching Birds*, his new list included policemen, postmen, naval officers, aircrews, financiers, writers, poets, teachers, artists, journalists, doctors, dentists, chemists, BBC producers, librarians, and clergymen.[82]

Fisher's natural history of the fulmar – with its mix of maps, charts, and drawings, and its solid exegesis on the biology of the fulmar – was a model of the new ornithology. Yet as its opening paragraphs revealed, it offered the reader more than dry science. Fisher's skill as a communicator was to engage his readers with the drama of the fulmar's spread. The book was also notable for the way it drew into his account of the life history of the fulmar an older bird-centred culture. This was the tradition of harvesting fulmars, or fulmar fowling, that took place on the remote island of St Kilda. In a long, diversionary chapter of the book, Fisher was drawn onto the terrain of social anthropology, detailing the culture of bird harvesting and its place within St Kildan society prior to the evacuation of the islands' human inhabitants in 1930. He combined this social study with a reassuringly quantitative assessment of the numbers of birds taken on St Kilda since the 1830s.[83]

Fisher's analysis of the fulmar fowling on St Kilda was notably shorn of any critical reflection about the practice of harvesting wild birds.[84] Rather, it was a sympathetic view of a society totally structured around its reliance on seabirds. The chapter ended with Fisher being drawn into a moment of imaginative reverie for a way of life that no longer existed and for the landscape of St Kilda and its ghostly birds. He noted how visitors to the islands in the nineteenth century had regularly recorded how the houses of the St Kildans were pervaded by the strong smell of fulmar. The 'ghost of this smell' came back to Fisher during a visit in 1948. He recalled how the clothing of bird ringers was soiled with the 'oily vomit' of fulmar chicks, and as their clothes dried before the fire the smell of St Kilda's bird fowling past came to life again. The smell thereafter formed an indelible link for Fisher with the fulmar and the dramatic landscapes it inhabited. As he put it: 'I can never smell that smell … without dreaming myself back on to Conachair or the Cambir, at the hard steep edge of Europe, among the fulmars.'[85]

This romantic, sentimental reverie chimed with how Fisher framed his whole study. As the quote above, taken from the opening lines of *The Fulmar*, reveals, the analytical science of the fulmar census and the biological account of the bird were overlaid with an obsession with the bird and its landscapes. In these opening, confessional passages, Fisher revealed how the bird had gripped his imagination, provoking quests to remote islands and across unpredictable cold seas. For Fisher, the fulmar had the power not only to animate and enchant the landscapes and seascapes it inhabited, but also to lure him onwards into his adventurous study. This was a seduction unwittingly performed by what he called the bird with 'cold dark artic eyes'. As if to undercut this drift into sentimentality, and as a good 'new ornithologist', Fisher also pointedly compared the birds to modern aircraft. Awed by their stiff-winged flying prowess, he suggested that fulmars were like high-performance 'sailplanes'. Compared with them, other seabirds like gulls looked clumsy in their flight action – 'trainer aircraft' to the fulmar's 'sailplane'. No wonder, Fisher concluded, that there was a connection between 'the Coastal Command aviator and the sailplane enthusiast ... and the finest natural sail-plane in the North Atlantic'.[86]

THE REDSTART AND PRISON BIRDING

In your prison camp you will no doubt, as an officer, have some if not much leisure in which to think and read and write. You will be as enisled as we are [on Skokholm] and your observation will be limited to the fauna of that entity of your camp and its perimeter. This should give you the chance of concentrating particularly on one or two species as I have done here and making a study of great interest.

From a letter to John Buxton from Ronald Lockley[87]

James Fisher's enthusiasm for the rugged, remote islands along Britain's northern and western coasts was shared by other prominent figures committed to the new ornithology. These other 'island lovers' included George Waterston, who had worked with Fisher on the fulmar study and was central to the creation of the Fair Isle bird observatory in 1948, and Peter Conder, warden of Skokholm Bird Observatory in the late 1940s and secretary of the RSPB in the 1960s and 1970s. Prominent too were Ronald Lockley and John Buxton. Lockley established Britain's first bird observatory on the island of Skokholm off the Pembrokeshire coast in 1933. He had taken out a lease on the island in 1927, attempting to farm it and study its bird life. Like St Kilda for Fisher, Lockley saw Skokholm as an 'island laboratory', set apart from human interference.[88] The island was both an escape from the modern world – a Rousseauean 'Eden away from Society' – and a place where the most modern techniques of bird migration study and research on individual bird species could be undertaken.[89] Lockley's pioneering migration studies, including his building of the first Heligoland trap in Britain, attracted leading ornithologists like Huxley, Fisher, and Max Nicholson as visitors to the island. Lockley worked with Huxley on his short documentary film *Private Life of the Gannet* (1937), much of which was filmed on the island of Grassholm close to Skokholm. The film opened with an aerial shot of Grassholm and its gannet colony, depicting it as an isolated 'Eden' and 'island laboratory' for the study of the breeding behaviour of gannets. Fisher also enlisted Lockley's assistance for his own study of gannets and corresponded with Lockley about renting his house on Skokholm for a holiday with Huxley in June 1944.[90]

Lockley was also a mentor for John Buxton. Buxton had first visited Skokholm in 1937, and, as Derek Niemann suggests, he had quickly fallen under the spell of both the island and Lockley. Buxton married Lockley's sister, Marjorie, and stood in as warden on Skokholm while Lockley carried out migration studies on shearwaters around Portugal and its Atlantic islands in 1939. During his tenure on the island, Lockley encouraged Buxton to make his own single-species studies, just as he would do two years later when Buxton was a prisoner of war in Germany.[91]

After his time as warden of Skokholm, Buxton joined the 1st Independent Company (commandoes) as a second lieutenant in 1939. Buxton's war proved to be a short one, though: he was captured in Norway in 1940 and held for the remainder of the conflict in a variety of prisoner of war camps, including Oflag VII B in the Eichstatt region of Germany. James Fisher sent him a copy of his gannet paper while Buxton was held at Oflag, obtaining his address from Lockley.[92] Lockley corresponded with Buxton too, pressing him to use his 'enislement' in the camp to conduct a single-species bird study.

Armed with the observational and recording techniques of the new ornithology, but without any binoculars in the camp, Buxton set about a study of the birds. In this he enlisted the services of other birdwatching prisoners, including George Waterston, captured in Crete in 1941, and Peter Conder, captured in France in 1940.[93] Together they produced a study of 'Migration at Warburg in 1942' (their first camp) and a report on the 'Birds of the Eichstatt District from September 1942 to April 1945'.[94] The studies counted bird species, used bar charts to represent changes in bird numbers, and recorded the dates, weather, and direction of the birds flying over the camp. The prisoner-ornithologists also put up nest boxes on trees around Oflag VII B to study breeding birds. Buxton saw the boxes as offering a way of studying 'bird populations in the camp',[95] and they became central to the single-species study that he pursued while in prison. This focused on a bird called the redstart, in part because they were easily identifiable and relatively approachable, 'like robins at home'.

He wrote up his study in the prison camp in a school exercise book, and it was published, barely altered, as a New Naturalist monograph in 1950. In their preface to the book, the editors saw Buxton's work as exemplifying the ethos of the 'new naturalists', noting, with a certain unintended irony, that the discovering of new facts was the 'province of the amateur naturalist with some leisure'. This leisure was best used in the study of a single species, as Buxton had done.[96] *The Redstart*, as Buxton made clear, was the product of cooperative observation. Drawing on the assistance of other prisoners, he was able to gather a mass of notes covering more than 850 hours of observation of one pair of birds between April and June of 1943.[97] Though Buxton was a poet and tutor in English literature at New College, Oxford by the time the book was published, his monograph on the redstart was rigorous in its scientific account of the reproductive behaviour of the birds. It covered the establishment of territory, pair formation, the nest, egg laying and incubation, the care of the young, food, and migration. As Buxton noted, approvingly quoting an extract from *The Times* from April 1948, 'bird behavior had been removed from the irresponsible custody of poets and put in the safer hands of scientific ornithology and statisticians of avian affairs'.[98] In the chapter on pair formation, he cited the work of both Julian Huxley and David Lack (author of the life history of the robin) on bird behaviour. However, it is the weight of detailed field observations, lifted from the notes collected in Oflag VII B, that dominated the book. As Niemann has suggested, Buxton's descriptions of bird behaviour often had a 'graphic, filmic precision'.[99]

While *The Redstart* was rigorously empirical, building up more and more detail of behaviour, it occasionally revealed Buxton's creative imagination too. There are flashes of this in the more subjective and emotional responses to the birds that occasionally surface in his determined building up of the empirical evidence. In the introduction, for example, he concedes that he had chosen to study redstarts for more than practical matters of convenience. They were birds, he confessed, of 'grace and beauty', selected as well 'for the sweet gentle charm of [their] song'.[100] Buxton also emphasized that his study was the result of encounters with a unique group of birds, seeing his book as 'not a history of a species, but a selection made from the biographies of a few individuals'.[101] For Buxton, then, *The Redstart* was the record of a brief, but intimate, relationship with four pairs of redstarts – a relationship that captured a short moment in their lives and which had been played out within the precise geography of two prison camps. As Buxton put it, introducing his redstarts and giving the weight of uniqueness to their lives and his account of them: 'This study … is principally concerned with the lives of 4 pairs of redstarts. 3 of these lived in 1941 near the left bank of the river Schzach where it passes the former Archbishop's Palace at Laufen in Bavaria and the other pair lived in 1943 near the left bank of the small river Altmichl where it flows through the former valley of the Danube a little below Eichstatt, also in Bavaria.'[102]

Buxton's emphasis on the unique individual lives of his birds rubbed up against the concern with bird populations and the turning of the living bird into abstract data within the new ornithology. His book was avowedly about unique individual birds and a geographically and temporally bounded moment in their lives. Buxton was also moved in his observations by the apparent indifference of the birds to the attention focused upon them. They lived, enchantingly, 'in another world than I', caught up in the immediacy of their own life world. This was not just Buxton the prisoner envying the birds their freedom to move in and out of the perimeter of the camp. He was also moved by the fact that they 'lived wholly …to themselves', living only in the moment, 'unconcerned with our fatuous politics'. Ultimately, it was their otherness from human life that drew his affection, even more so than the scientific value of his observations. As he concluded early on in the book: 'I loved them for their own sake and not for the sake of adding to men's knowledge.'[103]

BIRDS OF THE HIGHLANDS

> I knew [in 1933] that a new type of bird-watcher was emerging, one who sought to know a lot about a little rather than a little about a lot, and that he must be content to live for years in lonely places far away from the comforts of civilization while recording painstakingly all the time
>
> Desmond Nethersole-Thompson, *The Greenshank* (1951, 3)

> I do not regret that I first learnt to watch birds and to know birds as an egg collector.
>
> Desmond Nethersole-Thompson, *The Snow Bunting* (1966, 262)

On 29 March 1934 Desmond Nethersole-Thompson, history and classics teacher, departed King's Cross station for the Scottish Highlands. It was the start of an enduring relationship with the landscape, politics, and society of the Highlands, particularly Strathspey and the high tops of the Cairngorms. Out of this relationship would come a series of studies of the birds characteristic of the region, including four species monographs and other publications on the birds and natural history of the Highlands. The first, and the best regarded, of these was his study of the greenshank, which was the fifth species monograph published in the New Naturalist series.

As the editors conceded in their preface to *The Greenshank*, Nethersole-Thompson had probably spent more hours in the field than any other naturalist in the New Naturalist series.[104] The book was the product of seventeen years of intensive watching, much of it spent living with his family in an isolated bothy in the foothills of the Cairngorms or camped out for weeks at a time on the high tops of the Cairngorm plateau.[105] Unlike the collective, corporate fieldwork that shaped the studies of Fisher and Buxton, Nethersole-Thompson's field research was rooted in his family life, with his children and, especially, his wife, Carrie, playing a central role. Those who observed Nethersole-Thompson close at hand were struck by this distinctive kind of cooperative study. Nat Tracey, of the Association of Bird Watchers and Wardens, in correspondence about Nethersole-Thompson becoming an RSPB 'watcher' in Speyside, noted that: '[Nethersole-Thompson's] wife helps him considerably in his watching and often stays up on the hills with him for weeks at a time. She is very strong and is a very good naturalist.'[106] Nethersole-Thompson himself always emphasized the integral role played by his wife in the studies, crediting Carrie with particular skills as a nest finder and as an equal partner in their deep immersion in the study of the greenshank and other Highland birds. The couple's children were also co-opted to the research, with their first son Brock carried, as a baby, into the hills in a game sack alongside their equipment and provisions. Brock would later become an expert nest finder and participant in the family's studies.[107]

The Greenshank is a highly personal book in which the 'adventure in simple living' that Nethersole-Thompson shared with his wife and family underpins his account. On occasions this has an absurd, comedic quality, like when he describes how one pair of greenshanks vigorously attacked Carrie whenever she wore her Inverness cape.[108] More often, though, it is the shared, intimate encounters with the birds in the land that the Nethersole-Thompsons shared with them that surface in the narrative. As he notes early on in the book:

'Carrie and I have lived most of the time [in the seventeen years of intensive watching] ... among the foothills of the Cairngorms... We have known greenshanks in their many moods. We have reveled in the wonderful forms of an elaborate and colourful courtship and begun to understand the intimacies of their sex life... We have enjoyed their company amid the grandeur of loch and mountain in Strathspey.'[109]

This passage is telling for the way it meshes a sense of the shared life of the Nethersole-Thompsons with the domestic life of the greenshanks, seeming almost to slip between the human and avian forms of partnership and the building of a shared life together, for the Nethersole-Thompsons as much as for the birds. The Nethersole-Thompsons' intense study of and fascination with the birds is matched by their love of their shared homeland. As they expressed it in a rare joint article about their research in *Country Life* in 1940: 'He who seeks the greenshank's secrets will never regret his choice. His quest will lead him to the sour-peat bogs and stony grandeur of glen and mountains in Strathspey.'[110] Across Desmond Nethersole-Thompson's writing there is a recurring love expressed for Strathspey and the wider Cairngorm plateau. In both *The Greenshank* and the article co-authored with his wife, he also used very specific places and landscapes within Strathspey to set the drama of the stories of his greenshanks. One of these micro-geographies was a 'vast natural amphitheatre' in which the 'love-making of two greenshanks was closely and delightfully observed'. The field study was also shaped by a seasonal rhythm: 'In the snowbound corries ... eagles are already breeding when tarn and glen ring again with the music of the homecoming greenshanks. Thenceforward, throughout spring and summer, in weather fair and foul, one must unremittingly watch, study and record each phase of a brilliant pageant.'[111]

In *The Greenshank*, Nethersole-Thompson combined elements of the 'pageant' of the lives of the birds with detailed records of population density, territory, and courtship, together with other aspects of breeding biology and a speculative chapter on 'psychology and emotion' of the inner mind of the birds. Tables documenting the number of nests and the types of habitat, the occupation of territories, the times of laying, clutch sizes, and population densities were presented, together with a bar chart documenting the number of breeding female greenshanks in Strathspey.[112] This attention to counting and measuring informs Nethersole-Thompson's concern with grasping elements of the breeding ecology of greenshanks and compliments his interest in theorizing about the behaviour of the birds. In this latter enterprise, Nethersole-Thompson's work owed much – as he acknowledged – to his precursors among the 'wader watchers' such as Julian Huxley, Niko Tinbergen, and G.F. Makkink.[113]

Nethersole-Thompson's commitment to the best insights of ethological and ecological thinking and to a systematic, scientific presentation of evidence co-existed in *The Greenshank*, and in his other studies, with a disclosure of his emotional feelings towards the birds. As he insisted: 'We gradually came to regard our greenshanks not so much in terms of dry statistics of a population study but as old friends with whom we yearly renew our acquaintance.'[114] This sense of friendship and familiarity with the birds stemmed from the sustained and enduring nature of the fieldwork. Like John Buxton, Nethersole-Thompson sought to grasp the individuality of the birds he studied. He did so in a more dramatic way than Buxton by giving names to his birds, with the result that we are presented with a cast of what can often read like a soap opera in the narrative of *The Greenshank*: 'Old Glory, Jill, Glorious, Otter, Elizabeth, Fanny and Elizabeth, the wives of George … the two great ladies of our greenshank history'.[115] This candour about his closeness to and feelings for the birds was also evident in Nethersole-Thompson's physical handling of them. He regularly stroked 'Old Glory', 'our favourite hen greenshank', while she was incubating on the nest. In one of the few colour photographs in the book, Nethersole-Thompson is shown, in his army uniform, placing his hand on the back of 'Old Glory' while she sits on the nest. This 'petting' of a wild bird revealed his desire for a physical, and not simply an observational, connection with it and is a sign of his affection for 'Old Glory'. Physical touch – crossing the boundary separating bodies – effects an intense connection in this human–avian relationship.[116]

In developing this close relationship with named, individual birds, the Nethersole-Thompsons had to be able to identify their birds. This was no easy task since the individuals of each sex were difficult to separate by sight. Researchers studying bird species through the twentieth century developed various way of marking individual birds to overcome this general problem. This ranged from the use of coloured dyes to the attaching of rings to birds' legs. Nethersole-Thompson's solution was different and revealing. He drew on the knowledge of the oologists that the eggs of particular individual femals had the same shape and colour pattern. This enabled him to identify individual females by their eggs. 'Glorious,' he confessed, 'laid some of the most beautiful eggs I have ever seen.' Elizabeth's eggs were pictured in colour in the book, displayed in tabular form against a white background. They had been laid, and collected, between May 1921 and May 1933 by E.P. Chance before being donated to the British Museum (Natural History).[117]

Nethersole-Thompson's familiarity with oological wisdom and his comments on the beauty of the eggs were not accidental. He had been a dedicated egg collector in his childhood and his early adult life, and he was well known within oological circles in the 1930s and 1940s. He was mentored in the art of nest finding by the leading collector-ornithologists of the first half of the twentieth century, including John Walpole-Bond and Francis Jourdain. Jourdain was the most celebrated of the oologists who sought to link egg collecting with field studies, emphasizing its connection with modern scientific ornithology. He was the co-author, with H.F. Witherby, Bernard Tucker, and Norman Ticehurst, of the definitive five-volume book *The Handbook of British Birds*, which was published between 1938 and 1941. The book incorporated some of the thinking of the new ornithology, including work on bird territory and behaviour. It also contained plates of birds' eggs.

Following Nethersole-Thompson's move to the Highlands, he corresponded with Jourdain on a weekly basis, and in the late 1960s Nethersole-Thompson still affectionately recalled the influence on him, and upon scientific ornithology, of the man he called 'J'.[118] He even revealed that it was Jourdain who had introduced him to the work of ethologist Niko Tinbergen in 1935.[119] In *The Greenshank*, Nethersole-Thompson was more circumspect about declaring his loyalty to Jourdain and the other collectors, but he did locate his own study in a longer tradition of natural history expeditions to the Highlands by bird collectors. This may have been a bloody history, but he conceded that there was continuity between his studies and those of his precursors as much as there was a break with this past. As he put it: 'For the greenshank … the nineteenth and early twentieth century were a period of unending and merciless persecution… The saga of the greenshank was thus punctuated by the reports of shot guns, the splash and patter of water dogs and the weird music of drill and blow pipe in hotel bedrooms. The story I tell is rather different. There has been hunting in it, but it has always been with an aim and purpose in view.'[120]

One important element of continuity between Nethersole-Thompson's bird study and the practices of the collectors was the search for nests. *The Greenshank* is a study of the breeding biology and reproductive behaviour of the bird, and as such it depended on the finding and watching of nests. The book is also illustrated with many photographs (by Eric Hosking) of the birds on and around the nest site. Nethersole-Thompson acknowledged that he had found nest hunting a 'great challenge' since his schooldays.[121] The pleasure of this sport continued to shape his field craft. In *The Greenshank* he presented nest finding as a 'duel of wits' with the birds. In his last book, *Waders*, co-authored with his daughter Maimie in 1986, he continued to assert that 'nest finding is a field sport which is the first step to greater knowledge'.[122]

Nethersole-Thompson's relationship with the sport of nest finding and egg collecting, however, did not persist only in a reconfigured form in his bird studies: it was a sport he found hard to let go of. Following her father's death, Maimie revealed in the 1990s that he had not finally stopped collecting until the 1960s.[123] Following his move to the Highlands in the mid-1930s he continued to collect, combining it with his new studies. This brought him into conflict with the bird protectionists. In a letter to Preston Donaldson, then the RSPB's president, Nat Tracy, honorary secretary of the Association of Bird Watchers and Wardens, suggested that Nethersole-Thompson was subsidizing his research by finding the nests of rare birds and taking collectors to them.[124] Nethersole-Thompson, through the auspices of Eric Hosking, the photographer whom he had worked with in producing his co-authored article on greenshanks for *Country Life*, suggested to Tracy that he could be persuaded to give up egg collecting if the RSPB paid him to be a watcher in Speyside. Encouraged by the RSPB hierarchy, Tracy convened a meeting with Nethersole-Thompson, Hosking, Brian Vasey-Fitzgerald (the editor of *The Field*), and the oologist D.W. Musselwhite. Thompson agreed to give up collecting and keep his former clients away from Speyside. He would also use his influence within the British Oologists Association to ask them not to oppose the Wild Birds Protection Bill being brought forward by the RSPB. In return, Thompson wanted 15–20 collectors to be granted licences to collect eggs and to be paid to protect rare birds in the Spey Valley.[125]

Julian Huxley wrote to the chairman of the RSPB to lend his support to the plan, and Nethersole-Thompson persuaded Francis Jourdain to back it as well, though Jourdain's death later in 1940 held up the discussions.[126] Some within the RSPB were troubled by this search for a 'gentlemen's agreement' with egg collectors. Nethersole-Thompson, in particular, remained a divisive figure. Geoffrey Dent, head of the RSPB's Watchers Committee, warned Donaldson, the society's president, that Thompson was 'an unreliable and unpleasant person… He has skimmed the Spey Valley of eagles and dotterel and might take our money and at the same time take egg collectors to other districts.' Despite his strong reservations, however, Dent conceded that 'it might be worth trying him, but he would need very close watching'.[127]

Nethersole-Thompson was appointed RSPB watcher in Speyside from 1941 and performed the role until 1953, with Carrie undertaking some of the watching while her husband was commissioned as an Irish volunteer in the British army between 1942 and 1945.[128] While he did use his influence to keep members of the BOA away from Speyside, his report for 1953 noted that there had been a number of 'raiding parties of collectors' in the valley.[129] Nethersole-Thompson himself was also drawn back to nest and egg hunting by the return of a rare bird of prey – the osprey – to the Spey Valley in the mid-1950s. Thompson confessed to the particular fascination that ospreys held for him and to his long-held ambition to own the eggs of Scottish ospreys. In his book *Highland Birds* he recalled the excitement of the ospreys breeding again in Speyside, describing the period as 'those James Bond days of the osprey's homecoming'.[130] This seemed to be an admission of irregular activity on his part, including his disturbance of a pair of breeding ospreys and the disappearance of their eggs in the mid-1950s. At the time, George Waterston, who was in charge of the RSPB's militarized operation to protect the ospreys in Speyside, accused Nethersole-Thompson and his son Brock of climbing the nest tree and either taking eggs or forcing the birds to desert their nest. For Waterston and his colleagues, Nethersole-Thompson was ' public enemy number one', and he had to be excluded from the tight circle of birdwatchers protecting the birds.[131] The antipathy between Nethersole-Thompson and other, more protectionist-orientated, 'new ornithologists' such as Waterston revealed the charged politics that birds' eggs could generate, even between those birdwatchers shaped by the new bird science. As we will see in the next chapter, birds' nests and eggs remained central to, but highly controversial within, post-war cultures of birdwatching and bird protection.

CONCLUSION

Looking back on the early years of the BTO as it celebrated its golden anniversary in 1983, Max Nicholson emphasized its success in mobilizing 'energetic and dedicated amateur birdwatchers' and in welding them together to pursue corporate 'projects that advanced the science of ornithology'. Through this, the BTO had become, he suggested, an important organizational home for those who wanted to spend their time 'in the field advancing knowledge of living birds'.[132] Nicholson's was a fuller and more eloquent justification for amateur mass observation of birds than had been given by Harrisson and Hollom in their notes on the great crested grebe census in the early 1930s. They had remarked that volunteering for the census would be a good hobby for 'those people who find life dull'.[133]

While being part of a new democratic science helped Harrisson and Hollom overcome the humdrum nature of modern life, for Nicholson it had a more compelling purpose. He saw the fieldwork of the BTO as redirecting the collecting urge felt by many of those interested in wild birds into more positive, collaborative contributions to ornithology. It was a view shared by others of his generation, for whom the destructive pastimes of collecting eggs and skins could be transformed into the useful science of counting, ringing, and studying.

In some of its communications and publicity, the BTO – and its key thinkers such as Nicholson – tended to emphasize the novelty and modernity of the new ornithology, separating themselves off from established scientific ornithology. Later historians and commentators on natural history and popular science have tended to entrench this perspective and cast the advocates of the new bird science as birdwatching moderns who broke radically from the past. As in other areas of social life, however, 'Victorianism' had a complex afterlife.[134] Modernity and tradition could co-exist, even within self-consciously-modern practices. When Max Nicholson paid tribute to Bernard Tucker – one of his allies in the promotion of the new ornithology who had died in 1950 – in the 1980s, he offered a revealing assessment of both his friend and the BTO. He suggested that Tucker embodied the mission of the BTO to 'bring together systematic museum ornithology and bird watching, academic study and field observation'.[135]

It has been a central contention of this chapter that if we are to understand the development of a new kind of birdwatching and the associated ways of writing about wild birds in the first half of the twentieth century, we need to grasp the way that older ornithological concerns and preoccupations around collecting, the sport of pursuit, and a focus on nests and breeding biology were shared by both the old and new science. As we have seen, the collection and display of birds' eggs were increasingly hard to reconcile with the new focus on the livng bird. What had made birds' eggs so compelling for the collector – namely, their beauty and variety, and the sport and adventure involved in their collection – had to be pushed to the margins of ornithology as a legitimate science and serious hobby. And yet, as we have seen, birds' nests – the repository of those eggs – remained central to bird study through its focus on breeding biology and the reproductive life of birds. The skills and sport of the collectors were transfigured into the 'sport' of data collection, including nest finding. As Max Nicholson had candidly asked in 1931, was not the 'new bird watching' either the 'most scientific of sports or the most sporting of sciences'?[136] As we saw in the career and writings of Desmond Nethersole-Thompson, the sport of nest finding, egg collecting, and field study were strongly intertwined. As he practised it, the 'new bird watching' did not break completely from the old.

While Nethersole-Thompson may have been an extreme example, his life and writings do speak to a broader phenomenon: the interfusing of the old and new bird science. This was certainly evident in the biography and career of Nethersole-Thompson's friend Derek Ratcliffe. Ratcliffe had been drawn to wild birds and birdwatching through his enthusiasm for egg collecting in the 1940s and 1950s.[137] He became a leading scientist at Nature Conservancy (NC) and led the BTO's survey of peregrines in the early 1960s (which I discuss in chapter 4). But like Nethersole-Thompson, Ratcliffe remained committed to the value of selective egg collecting, despite publically criticizing the hobby. In recruiting observers for the NC peregrine survey, Ratcliffe reached out to all 'ornithologists ... both oologists and protectionists', as he strikingly put it, in order to ensure that the survey was as comprehensive as possible. He also revealed his admiration for the knowledge of egg collectors about the distribution of peregrines, conceding that he was 'forced to admit that this fraternity probably knows more about the distribution of peregrines than any other body'. If the wholesale taking of eggs was 'deplorable', the knowledge of ooloigsts was also valuable. Tellingly, in a hand-written postscript added to a typed letter to the falconer Humphrey ap Evans, Ratcliffe added: 'The Trust would, I think, deny "official acceptance' of egg collectors, but they are there and we just cannot ignore them.'

Ratcliffe's openness to the oologists proved to be significant in his scientific research into the causes of the decline of peregrines that his study for the BTO revealed. Following an international conference organized by the American bird scientist Joe Hickey into the possible reasons for the decline of peregrine populations across North America and Western Europe, Ratcliffe wondered whether the examination of historical records of the falcons' eggs held in pubic and private collections might reveal evidence of egg shell thinning over time. Desmond Nethersole-Thompson helped Ratcliffe to negotiate with the private collectors and gain access to their illegal collections, 'opening up the doors', as Ratcliffe himself put it, of '[this] underworld'. By drawing on this extensive sample of eggs, Ratcliffe was able to show that there had been a significant decrease in the thickness of peregrine egg shells following the introduction of the insecticide dichlorodiphenyltrichloroethane (DDT) into the United Kingdom. This proved a crucial piece of evidence for Ratcliffe in establishing the causal link between organochlorine pesticides and the collapse of the breeding populations of peregrines in the United Kingdom. For Ratcliffe, as well as for Nethersole-Thompson, this evidence vindicated the value of the older culture of collecting and its link with the new science.[138]

The example of Nethersole-Thompson also tells us much about both the place of a landscape aesthetic in the new natural history writing and its blend of science and sentiment. As we saw, his account of the breeding biology of the greenshank was understood within the precise landscape of Speyside. For him, and for other new birdwatchers such as James Fisher, encounters with wild birds involved a distinctive emotional geography. More than that, the birds helped to animate and enchant the places in which they were encountered. To know the birds was not simply to learn about their habits but their habitation. It meant coming to know their life histories and their environments. This was an emergent ecological aesthetic and practice.

The place-specific forms of enchantment with wild birds formed one part of a wider emotional and aesthetic appreciation of them that was evident in the new natural history writing. As we have seen, writers such as Fisher, Buxton, and Nethersole-Thompson experienced strong emotional reactions to particular species of birds. This ranged from enchantment to an almost erotic seduction via aesthetic appreciation of their plumage and song. Nethersole-Thompson and Buxton came to know particular birds as individuals, forming intimate connections with them and taking vicarious enjoyment from their lives. These were passions for birds that formally overlaid the rationalist methods and techniques of the new ornithology. They were also feelings produced out of the sustained watching and observation of the living bird within the new bird science. These were undoubtedly emotional relations that were different from those expressed through the older cultures of collecting. As Mark Burrow poignantly observed of the Carolina parakeet – declared extinct in the United States in 1939 – little was known about its life and behaviour despite the immense number of specimens of the bird and its eggs held by collectors.[139] Collectors may have marvelled at the colour of its eggs and been attracted to its plumage, but they did not sustain an emotional connection with the living bird. Its extirpation meant later birdwatchers were also denied that possibility.

Writing about their feelings for wild birds was part of the literary genre of new naturalism for its authors. Not only did doing so help their readers to extract enjoyment from what were often rather dry exegeses on breeding biology, population, and numbers, but it was also a riposte to the critics of the new ornithology, who saw the new bird science as a cold and soulless practice, obsessed with the counting of birds. For all the early success of the New Naturalist series, however, its popularity began to wane by the mid-1960s. Even at the height of its success, few of the books sold more than 5,000 copies. What attracted readers were field guides rather than specialist species monographs. What is more, birdwatchers and bird lovers increasingly looked to colour magazines, films, and television documentaries for their natural history entertainment. Understanding how some of these cultural forms shaped popular engagement with wild birds, and their relationships with the bird protection societies that expanded in the late 1950s and 1960s, forms the focus of the book's next chapter.

2

A Nation of Birdwatchers: Bird Protection and Recreational Birdwatching

One or two bird-identification books in the living room, a bird table in the garden, a pair of binoculars hanging in the hall ready for the weekend – these are normal routine items in many, may homes.

Desmond Hawkins, *The BBC Naturalist* (1957, 8)

In 1954 the Royal Society for the Protection of Birds (RSPB) produced a short silent film called 'Bird Watching for Juniors'.[1] The film began with an instructive drama. This showed two young children – a boy and a girl – wandering along a country track in a flat rural landscape. Their walk, however, was not as innocent as it looked. When they spotted a bird on its nest in a clump of reeds, they leapt into action. Rushing towards the bird, they plunged their hands into its nest without hesitation and took out four small eggs from the cup-like structure. Their actions had not gone unnoticed, however. Not far away, an adult birdwatcher had seen their plundering. He apprehended the children and explained to them that they should not have taken the eggs, and he helped them put the eggs back into the nest. The adult birdwatcher then invited the children to enter his hide and observe another nest, this time without disturbing the birds. The viewer then saw what the children saw: extraordinary close-up views of an adult bird feeding its young. The children left the hide and greeted the birdwatcher again, enlightened in the ways of appreciating birds through observation rather than egg collecting.

From this educational drama, the film moved into a more conventional documentary format. It presented the opportunities that the RSPB's Junior Bird Recorders Club (JBRC) offered the adolescent naturalist, including the ability to participate in exciting outings to remote bird reserves. Across these scenes the film emphasized how these young bird recorders were engaged in the close field observation of birds: ways of looking guided by the disciplines of note taking and sketching. By learning these techniques and skills the young JBRC members were moulded into competent observers and could become part of a convivial culture of birdwatching. As such, they stood at some distance from the bird's-nesters depicted in the film's opening scenes. While the latter was ill informed, thoughtless, and childish, the junior bird recorder was a studious, serious, and cooperative individual attuned to the value of science. Being part of the JBRC was the means to developing this new approach to enjoying birds.

'Bird Watching for Juniors' was produced at the beginning of a moment of growth and development for the RSPB. Following a decline in its membership in the 1920s and 1930s in the wake of a successful campaign against the plumage trade, the Society had seen a steady increase in those joining through the late 1940s and 1950s.[2] When the Society marked its Diamond Jubilee in 1949, its membership stood at just under 6,000. By 1955 this had reached 7,000, and the growth accelerated further through the 1960s. During this decade alone the Society's membership increased fivefold, and between the beginning of the 1960s and 1980 member numbers soared from 10,000 to 300,000.[3]

The RSPB's phenomenal growth as a membership organization was fuelled by wider social changes in the immediate post-war decades. Increased affluence, greater mobility, and more leisure allowed larger numbers of people to devote time and money to outdoor recreations. From the late 1950s more people visited the countryside; enjoyed holidays in self-catering accommodation outside traditional holiday resorts; began to travel abroad (particularly in Europe); and spent more money on bird books, binoculars, outdoor clothing, and bird food.[4] In the process, the social profile of birdwatching – and the RSPB's membership – changed. It shifted from being a pastime dominated by the middle classes to one that was more socially diverse in its class composition. By the late 1970s the RSPB's membership had certainly shifted from being predominantly middle class to having a strongly lower-middle-class profile. This set the Society apart from other voluntary conservation bodies, which remained overwhelmingly middle class in their social make-up.[5]

As the RSPB grew it outstripped other bird organizations like the British Trust for Ornithology (BTO), which remained small in comparison. The BTO had slowly expanded from 2,000 members in the early 1950s to reach 5,000 by the early 1970s. By the end of the latter decade it had just over 7,000 members.[6] With much of its income deriving from government funding through Nature Conservancy and research council grants, the BTO was able to survive without having to expand its membership as a source of revenue.[7] Because of this it prioritized the provision of research services and pitched its appeal to a relatively small, scientifically committed membership.

The RSPB followed a different path, embracing the casual birdwatcher as much as the committed bird protectionist and student of bird study.[8] Its public profile as a national body also benefitted in the mid-1950s from the success of another campaign for law reform in which it played a leading role. This was the passing of the Protection of Birds Act (1954), a law that simplified and strengthened the plethora of existing, largely county-based, wild birds protection orders. The Act effectively gave all British birds, both resident and migratory, and their eggs and nests stronger protection under the law, save for a number of birds deemed pests, a group of common breeding birds, and those game birds that were dealt with under the Game Laws.

While it was the product of compromises between protectionists, sporting, and farming interests, the Protection of Birds Act represented what a later government subcommittee suggested was a 'signal advance', providing a new legal code and consolidating a shift in public opinion towards greater protection of birds and their eggs. The journal *British Birds* was even more effusive, seeing the Act as 'an advance of almost Utopian character'.[9]

If the passing of the 1954 Act buoyed the RSPB, it sought to develop further its educational message about bird protection and to promote the enjoyment of wild birds. Woven through these messages the Society championed the innovations in field ornithology and bird study pioneered in the inter-war years. As the film 'Bird Watching for Juniors' made clear, this emphasized the observation, recording, and studying of birds. This message was combined with an aggressive attack on established popular pastimes that threatened the welfare of wild birds. These included, centrally – as 'Bird Watching for Juniors' again emphasized – the pursuit of bird's-nesting or egg collecting. This hobby encompassed not just the relatively small number of dedicated adult collectors like those who subscribed to *The Oologists' Record* (the subject of most attention in the 1930s and 1940s from the new bird science), but a larger swathe of children and young people for whom bird's-nesting was a well-established and routine pursuit. Against what the RSPB caste as the childish and selfish bird's-nester and the dangerous adult egg collector, it aimed to shape a forward-looking birdwatcher who was attracted to bird study, identification, and protection.

In this chapter I explore the role played by the RSPB in shaping the new, post-war birdwatcher, detailing the Society's moves to forge an affiliation between bird protection and birdwatching. In doing so, the chapter documents how the RSPB sought to shape the conduct of the 'new birdwatchers', organizing their feelings for birds into a distinctive recreational culture. In the first section, the chapter explores the moves by the RSPB to bolster birdwatching as a legitimate pastime by challenging the continuing popularity of egg collecting and bird's-nesting. In seeking to steer bird's-nesters towards the modern hobby of birdwatching and the forging of new kinds of attachments with wild birds, the RSPB had to confront the persistence of this established nature culture. This forced it not only to offer a progressive alternative to the pleasures of bird's-nesting, but also to bolster its pursuit of law reform and enforcement. The investigative work of the RSPB and its educational campaigns found support in popular books and films, but the Society encountered more opposition from policymakers, particularly within Parliament.[10] A significant number of MPs, in particular, sought to defend bird's-nesting and prevent the criminalization of what they termed the 'schoolboy bird nester'. I explore in some detail the parliamentary debates around the Act and consider what they reveal about the persistence of a different nature culture centred upon wild birds. This residual attachment to the pleasures of bird's-nesting among policymakers and a wider swathe of British society helped to prolong the hobby of bird's-nesting well into the 1970s, but in a moral climate increasingly inimical to it.

In the chapter's second section I consider how the RSPB sought to foster knowledge and enjoyment of wild birds through its publications and films, drawing out the way the Society institutionalized key elements of the 'new ornithology' within its educational message. The RSPB's promotional and educational work was supported by its film unit and by television's embrace of natural history, and by the development of new identification field guides and books that disseminated knowledge to an increasingly mobile population about where to watch birds and also how to enjoy encounters with them. These cultural forms helped to shape the growing popularity of watching birds. The popularization of the hobby generated, however, divisions between, on the one hand, bird protectionists like the RSPB and the promoters of bird science like the BTO and, on the other, supporters of 'rarity chasing' and 'listing'. The section reflects on the competing versions of the new birdwatcher, the moral geography of post-war birdwatching, and the charged ethics of the hobby.

EGG COLLECTING AND BIRD'S-NESTING

You will never stop small boys bird's-nesting… When I was a small boy I collected birds' eggs. I stopped when I was sixteen, which I think is what most of us do… I believe for small children to make a collection of eggs is [usually] the start of a lifelong interest in the subject, and a lifelong interest in the protection and care of birds.

Lord Tweedsmuir (1954)[11]

Settling the legal status of the practice of collecting wild birds' eggs, including for food, had proved to be a difficult problem for Parliament since it first attempted to legislate on the matter in the late nineteenth century.[12] In 1921 the Conservative government established a Wild Birds Advisory Committee to look at the issue of birds' eggs and nests as part of a broader investigation into the legal protection of wild birds. With little to show for the committee's deliberations over more than a decade, the RSPB pushed for the setting up of another committee of enquiry in 1938.[13] With the RSPB's own draft plans for new legislation and the work of Parliament interrupted by World War II, it was not until 1948 that a new Wild Birds Advisory Committee was established. Its deliberations between 1948 and 1953 laid the basis for the legal reforms that were enacted in the Protection of Birds Act (1954) – the landmark piece of legislation that shaped the treatment of wild birds through the post-war period.

In the parliamentary debates around the Protection of Birds Act two contradictory aims animated the interventions of many MPs and those in the Upper House: a desire to strengthen bird protection on the one hand, and the ambition to recognize the legitimacy of childhood bird's-nesting on the other. Many parliamentarians in both houses confessed to their childhood enthusiasm for the hobby and defended it as having shaped a healthy and lifelong interest in wild birds and natural history.

Their discussions crystallized around clause 2(4) of the bill, which gave the Home Secretary powers to draw up a list of common birds whose eggs were excluded from the protection offered in the bill. Sir David Maxwell Fyffe, the Home Secretary, had insisted on this clause, much to the consternation of the chair of the Wild Birds Advisory Committee, Lord Ilchester, and many of its members, who had argued for near-universal protection. In moving the second reading of the bill in the House of Commons, Lady Tweedsmuir called clause 2(4) the 'bird's-nesting clause'.[14] It was designed to prevent the criminalizing of children who 'bird's-nested'. As she argued: 'Children do not as a rule come upon the rare species, and they take a great deal of delight in bird's-nesting. I do not think the House would like to bar what is not only a most exciting pleasure, but one which gives lasting interest in wild life. I well remember how thrilled I was in my own small egg collection.'[15]

While Lady Tweedsmuir conceded that the clause could open the way for 'the hooligan and his wanton destruction' in taking common birds' eggs, she felt that this problem could be dealt with by education. As she suggested, 'a great deal has been done already by parents and schools to try to encourage the adventure of finding the nest but leaving the egg'.[16]

These views were supported by others. Major Tufton Beamish, MP for Lewes and a member of the RSPB council, who had played a key role in the introduction of the bill, emphasized the 'absurdity' of making a law that would mean 'every child who went bird nesting and took an egg should be liable to a penalty'. The 'destructive' and 'unscientific' practice of egg collecting should be tackled by education when it involved children. Professional dealers in birds' eggs, on the other hand, were a serious problem. Major Beamish revealed that he had known of a dealer who had sold between 20,000 and 25,000 lapwings' eggs in one year, the majority to children for 6d. The bill, even with clause 2, would prevent this outrage from happening.[17]

The government's position on the 'bird's-nesting clause' was also strongly defended by Sir Hugh Lucas Tooth, joint under-secretary at the Home Office. Having explained how the bill sought to deal with the protection of birds along 'national lines' and on a 'national basis', he set out three objections to the prohibition of bird's-nesting. These were, firstly, that it was a prohibition that would apply largely to children and was thus hard to justify; secondly, that public opinion would most likely view the outlawing of bird's-nesting as 'unnecessary and crankish in itself'; and thirdly, that it was not right in principle to make bird's-nesting a crime. The real threats to wild birds' eggs came from the 'depredations' of the 'collectors of rarities' and the destructive behaviour of 'young hooligans'. Bird's-nesting was, compared with these forms of behaviour, 'harmless'.[18]

When the bill was sent to the House of Lords in April 1954 the 'bird's-nesting clause' met stronger opposition. The Earl of Ilchester feared that the clause had effectively 'wrecked' the bill. The list of common birds' eggs that the Home Secretary was empowered to establish would open up, the Earl argued, 'an extraordinarily easy time for the people we want to stop – that is, the stealers of rare birds' eggs. It can become a paradise for them.'[19] Lord Hurcombe, later to become chair of the RSPB Council and the Society's president, also challenged clause 2, seeing it as a loophole through which 'a much more serious kind of collecting will go on'. He warned Their Lordships that they were probably unaware of the extent of this 'mania of collecting eggs' among adults and the danger they posed to scarce as much as rare birds.[20] Earl Jowitt, while sympathetic to Ilcheter and Hurcombe and committed to 'knocking out adult egg collectors altogether', defended the enjoyment bird's-nesting gave children. Rather than criminalize them, Jowitt urged teachers and others to encourage children to leave the eggs alone and learn 'to take delight in watching birds and get to know the different types'.[21]

The Protection of Birds Act received its Royal Ascent in December 1954, complete with the 'bird's-nesting clause'. As his critics feared, the Home Secretary moved swiftly to establish an order removing the eggs and nests of thirteen common species from the legislation.[22] The RSPB was particularly frustrated by the Home Secretary's decision and published a leaflet titled 'Schoolboy Birds' Nesting'. It claimed that the order 'seriously weakened the administration of the whole act' and had done so based on the false assumption that the police would actively prosecute young children. Even if they did, the RSPB felt that it was highly unlikely that magistrates would seek to convict. Moreover, the RSPB accused the Home Secretary of failing to recognize that 'the whole trend of modern bird watching [placed emphasis] on the study of living birds'.[23] This was a view reaffirmed by Peter Conder, the RSPB's assistant secretary, in a reply to a letter on the subject of the Common Birds Order from an RSPB member. As he wrote: 'We most strongly oppose [the order]... It is upsetting the efforts made by schoolteachers ... and ourselves to educate children into new ways of looking at birds.'[24]

The Home Office was unmoved by these objections. However, the ongoing pressure from protectionists and the exceptionally cold winter of 1962/1963 combined to eventually produce some movement within Whitehall. The new Home Secretary, Henry Brooke, stirred by evidence from the RSPB and an appeal from Lord Hurcombe in the House of Lords, reversed his predecessor's decision and added thirteen common birds to the list of protected nests and eggs. He had been persuaded that the extreme winter weather had taken a heavily toll on these species and that they needed new protection to ensure that their populations recovered.[25] Four years later, in 1967, a new Protection of Birds Act was passed, amending the 1954 Act and crucially removing the ability of the Home Secretary to introduce a Common Birds Order. Thirteen years after it was first introduced, the Protection of Birds Act gave legal protection to all British birds' eggs and nests, save for those that continued to be deemed pests and those covered by the Game Laws.

Between the passing of the 1954 Act and its successor Act of 1967, the RSPB sought to use the law to pursue commercial dealers and private collectors, establishing a subcommittee to investigate and prosecute the private trade in eggs. With a certain boyish enthusiasm and bravado, its members referred to this as the 'Dick Barton' subcommittee, named after the popular radio 'special agent'. The subcommittee followed evidence supplied by RSPB members, who were assiduous in reporting breaches of the 1954 Act, as well as briefly hiring a private detective to watch suspected egg dealers.[26]

The RSPB had some success in the early years of the Act in working with the police to prosecute dealers and the collectors they dealt with. Their prosecutions revealed the scale of commercial operations, with dealers' price lists detailing an enormous number of British and European species whose eggs could be bought, either as singles or clutches, and the variation in price. One price list included the following: 'blackbird, British song thrush, House sparrow, 1d. each; avocet 2/6; bittern 5/1, spotted eagle 15/1; golden eagle, 40/- to 60/- depending on markings'. Alongside the eggs were prices for all the accompanying paraphernalia of the hobby, including drills, blow pipes, egg solvent, and embryo hooks.[27] Philip Brown, the RSPB's secretary and a member of its subcommittee on egg collecting, worked with the authorities to check and inspect the eggs when police arrested collectors. This was no small task given the size of some of the private collections confiscated. These could run into thousands of eggs and, while they were usually meticulously labelled, Brown was required to check the identity of the eggs given the higher fines that existed for schedule 1 species (rare breeding birds).[28]

While it pursued and sought to assist in the prosecution of private collectors and dealers, the relatively small fines issued under the Act for the taking of proscribed eggs frustrated the RSPB.[29] The Society ran a regular column in its members' magazine that detailed prosecutions under the Act, and these reports developed a world-weary tone as the Society became increasingly critical of the weak enforcement of the law. The high cost of pursuing prosecutions and the limited nature of the penalties meant that the Society was forced to choose carefully those cases that it followed. Reviewing the RSPB's enforcement activity since the mid-1950s, Richard Porter, its 'Technical Officer (Bird Protection)', who had been appointed in 1968, suggested that 'the Society had a very patchy law enforcement history'.[30] Much of its work had involved passing on information to either the police or the RSPCA. Porter recommended additional staff, better liaison with the police, and engagement with field sports' associations and landowners. He also argued that, because of the high costs involved in pursuing prosecutions, only those 'especially bad cases and those where maximum publicity is likely to be forthcoming' should be taken up.[31] Porter's guidance referred not only to crimes associated with egg collecting, but also took in the shooting, poisoning, and trapping of birds of prey; the trapping and sale of wild birds; and the 'wanton shooting and destruction of birds (including nests and eggs)'. The crimes against birds of prey represented the biggest problem in the early 1970s, though egg collecting constituted the fourth highest recorded category of bird crime.

The RSPB's relative failure to check the practice of egg collecting was not only a result of its lack of resources and the relatively low status held by bird crime within policing and among the sentencing authorities, but also because of the continued defence of collecting and the way it was embedded within popular childhood cultures of nature, the latter facilitated by the continued publication of books on birds' eggs. Before and around the passage of the 1954 Act, *Country Life* ran a series of articles that, while conceding the need for effective bird protection, conjured the enduring appeal of birds' eggs and the thrill and adventure of childhood collecting. Most of these articles had a nostalgic and wistful air, with memories of childhood bird's-nesting serving to invoke happy recollections of school holidays.

Writing in August 1956, Geoffrey Grigson conceded that the new law was probably necessary but argued, like some of the parliamentarians in the debates in both Houses, that an enjoyment of nature had been fostered by the search for birds' eggs. Grigson also recapitulated the tantalizing appeal of birds' eggs. Looking back he recalled that they were a 'treasure surpassing the treasure of dwarfs, guarded by dragons'.[32]

An anonymous review of Richard Fitter's book *Pocket Guide to Nests and Eggs* in *Country Life* also invoked the powerful attraction of eggs. The recollections of the artist Paul Nash were quoted: 'Ever since I can remember … I have been delighted by the sight of birds' eggs. They represent for me a kind of beauty which to this day nothing supplants.' While the review noted how the hobby of bird's-nesting was now frowned upon, the lure of eggs was still there. Fitter's book recognized this but was careful, as the reviewer noted, to warn his readers to 'take the book to the nest, not the egg to the book'.[33] The book tellingly drew, however, on the published descriptions of eggs from the late F.C. R Jourdain, the leading oologist and ornithologist-collector, who had sought to reposition the illegal habit of collecting into the study of nests and eggs in situ. While the book was part of an attempt from some naturalists to reimagine the allure of eggs within the new legal framework and as part of legitimate observational practices, it nonetheless worked to affirm the beauty and desirability of eggs that bodies like the RSPB were keen to discourage.

While bird's-nesting appeared to be in decline as a pastime in the early 1950s, it remained a central part of rural childhoods and persisted within urban and suburban childhood culture into the 1970s.[34] Some sense of this culture was captured in a unique study undertaken by Mass Observation (MO) in 1951. The study sought to capture past involvement in bird's-nesting, current attitudes towards it, and its persistence among children. Nearly 350 panel members participated in the survey. It was supplemented with a questionnaire completed by 200 schoolchildren. Three generations were captured in these combined studies. The adults were divided more or less equally between those aged under thirty-five, who were children of the inter-war years, and those aged between thirty-six and forty-five, who were children of the Edwardian period. The schoolchildren had been born in the very late 1930s and 1940s.[35] Like other MO studies, the adult respondents were mostly educated beyond the age of sixteen and were more middle class than a representative sample of the population would have been.[36] Just under two-thirds were men.

The survey revealed that half of the respondents, both men and women, had gone bird's-nesting as schoolchildren, including nearly all those who had grown up in the country. Most had begun around the age of seven and had stopped by the time they were fourteen. Only a minority looked back with any regret on their childhood collecting, with fewer than one in ten of the men troubled by their past behaviour.[37] For many, bird's-nesting had been part of a broader childhood culture of games, collecting, and outdoor play and sport. One panelist, a thirty-five-year-old woman, remembered bird's-nesting as part of what she described as 'the collecting urge – at the same time I collected wild flowers and stamps and cigarette cards, as did my friends'.[38] Another, a twenty-seven-year-old man, saw bird's-nesting as a 'hobby ... more or less like stamp collecting'.[39] Many also recalled the thrill and excitement of looking for nests and the pride in their knowledge about birds and eggs. A forty-four-year-old women suggested her bird's-nesting stemmed from 'an affection for creatures, first of all; excitement of finding; pride in knowing something by myself; indescribable feeling of pleasure at seeing the nest, so intricate, so unbelievable'.[40]

It was clear from the respondents that the hobby of nest finding was governed by clear rules and ethics. Pre-eminent among these was the conviction that only one egg should be taken from a clutch. One panelist rationalized this by suggesting that birds would not miss a single egg since they could not count. Collecting was also justified as long as the bird did not forsake the nest. Another confessed that while they tried to stick to the one-egg rule, they sometimes took more from common birds' nests.[41]

The commitment of many of the panelists to these codes of honour was reinforced by their desire to emphasize that looking for nests had often led to a wider interest in birdwatching and bird study. It was sometimes associated with a normative judgement that children in the present should be steered towards birdwatching. A thirty-nine-year-old woman argued that children should be discouraged from 'actually removing eggs from nests, though not of course from indulging in the fascinating hobby of bird watching'.[42]

This concern to channel the collecting habit into birdwatching was more starkly expressed by the panelists who had never bird's-nested. A thirty-seven-year-old woman suggested that it was 'interesting to watch birds in all their activities, including building nests. But I don't look closely at nests for fear of disturbing the birds.' An eighteen-year-old man suggested that 'for anyone interested in natural history it would have a certain fascination and I cannot see anything wrong with it as long as they do not 1) harm or frighten the birds 2) take all the eggs 3) injure the nest'.[43] A thirty-nine-year-old woman who had loved birds as a child and still liked 'watching birds and … differentiating between them' felt bird's-nesting was 'wrong and would never encourage it in children… I think it is good to encourage bird watching.'

The moral outrage of those who had never bird's-nested was particularly striking in the study. They described bird's-nesting as 'senseless and destructive', 'absolutely horrible', 'cruel', and 'cruel and inhuman', and they talked of their 'revulsion' towards the practice.[44] A twenty-three-year-old mother was typical when she claimed that 'disturbing nests [was] extremely cruel and [I] will do my utmost to stop my child from disturbing nests'.[45]

The views of the 200 schoolchildren that formed the second strand of the study suggested that there was a continuing interest in the hobby, despite the broad consensus among the adult panelists that bird's-nesting should be discouraged. A rural correspondent from the Dorset/Hampshire border had suggested that in her area '"egg saving" and nest destroying … are still almost universal' and that 'destructive youths' still continued to take their toll on birds' nests, encouraged by the wider search for specimens to be added to many children's collections.[46] The draft report produced by MO set its face against this evidence and sought to draw a positive message from the study. It emphasized the 'whole trend … away from collecting … towards observing animals and birds'. This seemed to miss the continuing, if apparently declining, appeal of the hobby. Glossing over the continuing allure of birds' eggs, the report ended with a paean to the forward-looking and rewarding pleasures of bird-watching when set against the habit of bird's-nesting. As it argued: 'Bird watching … challenges a youngster's powers of observation, patience, skill in stalking and concealment and it can provide plenty of excitement too. Above all it is not a dead-end hobby like egg collecting; it provides unending and growing satisfaction which continues throughout life.'[47]

THE NEW BIRD RECORDERS AND WATCHERS

The conclusion of the MO report on bird's-nesting strongly echoed the arguments made by the RSPB. These were articulated not only in its policy statements and press releases, but also in its early films and in the pages of the *Junior Bird Watcher* (JBW), the club magazine for the RSPB's adolescent members. Launched in 1955, the magazine reinforced the appropriate mode of conduct and the ways of studying birds that were necessary to become a good birdwatcher. In earnest tones it rehearsed the tightly defined script for conducting bird observation and bird study, framed by the club's central message that the 'better protection of birds' came through 'studying their habits'.[48] Editorials and essays emphasized the importance of methodical observation, encouraging the magazine's readers to submit bird records to the club 'based on field notes' and to undertake small studies of familiar or common birds.[49]

If these prescriptions on how to be a good birdwatcher and student of bird study were allied to the club's mission to promote bird protection and to deflect young people away from egg collecting, they also sought to differentiate committed bird study from the 'mistaken' pursuit of 'rarity-hunting'.[50] As the RSPB's secretary Philip Brown warned in a preface to JBW, the 'methodical observation of a common species' formed the basis of the 'best field research' rather than the empty building of a 'long "life-tally" of rare birds'.[51] For the JBRC, then, the young bird recorder committed to bird protection was defined against the superficial tally-hunter. This was a theme, as we will see, that was strongly reaffirmed across the RSPB's other publications for its adult members through the 1960s and 1970s.

If the message of JBW was especially important for the RSPB, given its desire to draw adolescents away from bird's-nesting towards birdwatching, the Society's publications for its older members developed a similar conception of the enjoyment of birds through the modern techniques of birdwatching. On the pages of *Bird Notes* and *Birds*, in RSPB pamphlets, and in films and books sponsored by the Society, the new post-war birdwatcher was inducted into the disciplines of the hobby.

One early pamphlet produced in 1949, 'Bird Watching and Bird Recording' by P.H.T. Hartley, set out a rigorous model of conduct for RSPB members. It emphasized the links between birdwatching and bird recording, pressing its readers to be 'exact in observation, scrupulous in record, honest in interpretation and bold in speculation'.[52] Hartley practised what he preached, and his own diaries detail a more-or-less daily recording of birds around his Suffolk home and a longer study of the song thrush population in his garden and its environs.[53]

The RSPB's assistant secretary, Peter Conder, echoed Hartley's prescription for modern birdwatching, including its parochial focus, in *Bird Notes* in the early 1960s. He encouraged RSPB members to investigate bird populations within their own parishes and to consider conducting life history studies of common, local birds.[54] Conder's injunction was developed two years later in a special issue of *Bird Notes* devoted to 'bird gardening'. This encouraged readers to see their gardens as 'miniature sanctuaries' that could help and support bird populations under pressure from wider environmental change.[55]

An editorial in *Bird Notes* in the autumn of 1965 reiterated this message about conservation and garden bird study, encouraging members to initiate their own projects with garden birds. Giving the example of a study of birds' feeding preferences for different kinds of berries, the editorial sought to stir its readers into action, suggesting that '[these projects] will give bird lovers an added interest and the feeling that they are helping to provide useful information which might prove of scientific importance'.[56] It was striking how this idea of garden bird study spread into the wider culture. Advertising for Swoop (one of the first packaged food for wild birds) placed in the RSPB's members' magazine emphasized how the food not only attracted birds to your garden, but also was 'an excellent aid to bird study'.[57]

The RSPB's desire to promote this kind of active, study-orientated approach to enjoying birds continued in its new glossy members' magazine, *Birds*, launched in late 1966. The first issue ran a photo-essay on a water bird called the dipper based on the field observations of James Alder and his five-year study of the bird. Alder described his project as a life history of a number of dippers and

Figure 2.1 Junior birdwatchers (1963).

told of how he had built up more than 500 case histories. It was telling that the article was an abridged version of a long essay published in the journal *British Birds* three years earlier. *British Birds*, a key vehicle for the new bird science since the early years of the twentieth century, saw itself as a bridge between the BTO's serious survey and census work and the more casual birdwatching of RSPB members.[58]

While *Birds* undoubtedly drew on the new ornithology of *British Birds*, it combined this with the presentational advantages of colour photography and the glossy colour supplement format. The new representational possibilities opened up by the use of colour allowed the magazine to capture the beauty of birds and the landscapes they inhabited in rich colour tones. But *Birds* photojournalism avoided excessively aestheticized bird portraits and focused on the value of photography in capturing behaviour. Thus, while bird portraits did appear in the magazine, the emphasis was on using photography to record and illustrate studies of behaviour. In 1968, for example, *Birds* ran a feature on the nuptial display of the great crested grebe. Published to help promote the RSPB's film 'A

Waterbird's World', the article, like the film, focused on showing RSPB members some of the behaviour first described by Julian Huxley in the early years of the century. The article revealed that the improved recording techniques of filmmaking had made it possible for its director to reveal new dimensions of the 'weed ceremony' of the grebes not seen by Huxley.[59]

Starting in the mid-1950s, the RSPB's film unit produced around eighty minutes of bird documentary each year.[60] Early films, some made before the unit was formally established in 1953, focused on the Society's nature reserves, like Minsmere and Havergate Island, or nature regions like Orkney, Shetland, and the Scottish Highlands. The film of the latter region, 'Highland Birds', released in 1958, was one of the most ambitious of the RSPB's films to date and was later screened on the BBC.[61] The RSPB saw these films as having a 'propaganda value' in promoting the RSPB's aims and for recruiting new members. As Chris Mylne, the head of the film unit, expressed it in March 1957, the 'chief aims' of the films were to support the RSPB's ambition to educate public opinion 'on the aesthetic appeal of wild birds and on the need for their protection, on public support of the Wild Birds Protection Act and on the establishment and maintenance of reserves'.[62] Films could help the RSPB to achieve this mission, bringing to audiences 'aspects of birdlife normally beyond their reach and outside their experiences, with a strong aesthetic and emotional appeal'.[63]

The RSPB had first sought to reach the general public through film by joining with the Fauna Preservation Society to screen James Fisher and Roger Tory Peterson's documentary 'Wild America' at the Royal Festival Hall (RFH), on London's South Bank, in 1955.[64] The success of the event encouraged the RSPB to re-edit a series of its early films under the new title 'Birds in Britain', which it screened at the Royal Festival Hall, with an introduction by James Fisher, in January 1956. The RFH film shows became a central event in the RSPB's calendar from this point on, and they were central to its ambition to extend its public profile and to promote the cause of bird protection. The RFH screenings ran until the late 1980s.[65]

Alongside the annual public film shows in central London, the RSPB was keen to develop a national programme of film screenings. These 'screen tours' were envisaged by the Society as operating along the lines of the successful Audubon screen tours in America.[66] The aim was to provide more than 200 shows each winter, these giving the RSPB 'an absolutely first class "shop window"' to promote itself to the public and to potential new members. By the late 1970s, supported by the creation of local RSPB members' groups, RSPB film shows took place in town and cities across Britain, filling concert venues and halls and helping to weld together the associational culture of the Society's large membership.[67]

Many of the RSPB's films focused on aspects of bird behaviour, with attention given to the reproductive life of the birds.[68] These studies of behaviour typically brought a life history perspective to bear and detailed the problems and dangers the birds faced during the breeding season and on migration (if they were migrants).[69] For bird species that nested communally, like rooks, the films focused on the social life of birds and the social problems of bird society, including documenting disputes over status, property, and the place of violence.[70] Species-based films combined attention on behaviour with a strong landscape aesthetic. Images of sunsets over reedbeds, misty marshes, snow-covered fields with a church nestling at the edge of a wood, and snow-capped mountains placed British birds within particular nature regions and landscapes. In this regard, the films generated a powerful imagining of the British countryside and the connection between birds and a sense of place. This worked to position birds as an integral part of the natural and national heritage.

From the late 1950s, this landscape aesthetic was increasingly tempered by a sense of the threats facing the countryside and coast and the birds that formed part of these landscapes. As Chris Mylne suggested, it was important that the RSPB films covered these developments in order to build 'a climate of opinion where protection is seen in its right place as an essential need in an age of urbanization and mechanization'. The rapid pace of environmental change and a number of environmental catastrophes during the 1960s deepened this focus within RSPB films.[71] Films also documented the practical work that the Society did in protecting birds, regularly showcasing its reserves as spaces of protection and as locations where rare and scarce birds could be seen.[72]

The importance of the RSPB's film unit to its promotion of bird conservation and the enjoyment of wild birds ran alongside a recognition from within the Society of the increasingly important role that broadcasting played within both these endeavours. The RSPB had well-established contacts with the BBC in the early post-war years, and it sought to offer the corporation sound recordings for its radio series 'The Naturalist', one of three important BBC radio natural history programmes of the late 1940s and 1950s. These shows, aimed at amateur naturalists and a general audience, mixed studio content with field recordings of bird song and calls.[73] As the BBC developed its television service and new television natural history content, the RSPB continued to collaborate with it. Between 1957 and 1962 the RSPB's film unit provided a number of films for the BBC, including 'The Return of the Avocet' (1958), 'Scottish Highlands' (1958), and 'The Return of the Osprey' (1959).[74] The arrangement was important for both organizations. The BBC was keen to foster links with groups like the RSPB to allow it to access new wildlife films and to develop the network of wildlife filmmakers.[75]

The BBC's early natural history television included a series of monthly programmes with Peter Scott, founder of the Wildfowl Trust. The majority of these half-hour shows were built around films Scott had made of his expeditions but they also included filmed material from other naturalists. The most significant of these early contributions came from the Austrian filmmaker Heinz Sielmann. Scott had first seen Sielmann's film at the International Ornithological Congress in Basel, Switzerland. Broadcast on British television on 15 January 1955 and with commentary and narration by James Fisher, it showed footage of a family of black woodpeckers taken inside the nesting hole.[76] The film caused a sensation on its broadcast, with the BBC's audience research revealing viewer appreciation on a par with the coverage of the Queen's Coronation.[77] The film put the BBC's natural history television on the map and led to a regular and long-running series under the title of 'Look'.[78]

'Look' used a recreation of Peter Scott's studio at his home in Slimbridge as the set for the programme, mixing films, illustration, and conversation as part of a natural history lecture. 'Look' achieved viewing figures of between 6 million and 7 million, and it regularly scored very highly on the Reaction Indices used by the BBC to measure audience appreciation.[79] The BBC saw this genre of wildlife television as central to its 'professional view of natural history', which it aimed to develop through the 1960s. For senior BBC managers like Huw Weldon, the 'Look' films and talks carried more weight than the other genres of wildlife television also being produced by the corporation. These included the safari-adventure style films of Armand and Michaela Denis and shows like 'Zoo Quest', presented by David Attenborough, which recorded the collecting of exotic specimens for London Zoo.[80]

As with the RSPB's own films, there was growing interest within the BBC's natural history team on animal behaviour. This drew the BBC's Natural History Unit (NHU) into collaborations with the leading figures from animal behaviour studies, such as Konrad Lorenz and Niko Tinbergen. This attention to animal behaviour formed part of the NHU's commitment to giving 'the spirit of scientific enquiry … pride of place' in its filmmaking.[81]

This was evident in the programmes that took over from 'Look' in the late 1960s and 1970s, including BBC2's 'The World About Us' (1967–1987) and 'Wildlife on One' (1977–2005)[82]. The latter reached huge audiences in its first season, averaging 19.8 million viewers and eclipsing that mainstay of the television schedules, the 'Nine O'Clock News'. 'Life on Earth', the NHU's landmark thirteen-week series exploring the evolution of life on the planet, expanded the appeal of wildlife television even further, reaching audiences of 15 million and achieving record-breaking Reaction Indices on its transmission in 1979.[83]

During the 1980s the BBC continued to develop new natural history formats, including live outside broadcasts from bird reserves. These included 'Birdwatch from Slimbridge' (February 1980), 'Birdwatch from Minsmere' (1981), and 'Birdwatch live from Foulesleugh' (1982), the latter two reserves being managed by the RSPB.[84] The BBC's wildlife television output was crucial in helping to foster a popular, national enthusiasm for natural history. The RSPB benefitted from this, and its association with the BBC gave legitimacy to the cause of bird conservation and the gentle pleasures of birdwatching. Watching birds on television and film, however, was not the only feature of the leisure culture of birdwatching in this period. Print culture, especially books, was also central.

READING, BIRD STUDY, AND IDENTIFICATION

> To most of us bird watching is recreation – not only in the sense that it takes us out into the open air … but also because reading about birds, trying to identify them, watching their habits, gives us something to think about.
>
> Peter Conder, *Comment*, 'Bird Watching as Recreation'[85]

From the late 1940s and especially through the 1950s, 1960s, and 1970s, natural history publishing, and especially bird books, enjoyed a period of expansion and growth.[86] The growing market for books connected to birdwatching, including those supported by or published by the RSPB, was a key part of this phenomenon.

One of the most successful of the new swathe of books was *Book of British Birds*, a joint publication from the Automobile Association (AA) and Reader's Digest.[87] The book came with an RSPB membership leaflet inserted into it and was available to buy from the RSPB's sales department. It formed part of a series of books issued by Drive Publications that aimed to serve 'the motorist and his family' as they explored Britain's towns and countryside by car.[88] As an aid to a late 1960s version of the motoring pastoral, the book was directed less at the already committed birdwatcher and more at those families that had recently been attracted to the hobby. Drawing on the precepts of the new ornithology, the book mixed striking painted bird portraits set against white backgrounds, one species per page, with chapters on identification, bird flight and bird anatomy, theories of bird navigation and migration, territory, and behaviour, together with sections on 'birds and man'. The identification plates that preceded the species portraits were central to the knowledge and guidance the book offered its readers. As the title page for the identification section put it: 'An interest in birds begins with identification – with the pleasure to be had from being able to put the right name to a bird, whether studied at leisure in a town park or glimpsed fleetingly on a drive through the country.'[89] The last section of the book aimed to support the motorist birdwatcher in their pursuit of birds and was devoted to where to see birds. It closely followed the map-based format of James Fisher's 1966 book *Shell Nature Lovers' Atlas, England, Scotland, and Wales,* which was described by a reviewer in *Birds* as a 'handy pocket book for the mobile millions who drive through our countryside today in search of nature's wonders'.[90]

In his review of *Book of British Birds,* Peter Conder commended it for taking the reader beyond 'pure identification … and [giving] added interest in birds and their lives'.[91] The book repaid his good review: its inserted membership leaflet generated more than 7,000 new RSPB members in its first year of publication.[92] Notwithstanding the success of *Book of British Birds,* and despite Conder's comments, it was dedicated identification books or field guides that, since the mid-1950s, had come to increasingly capture the interest of the buyers of bird books. This new generation of guides drew strongly upon the new discoveries of 'field workers'. The best-selling and most influential was *A Field Guide to the Birds of Britain and Europe* by Roger Tory Peterson, Guy Mountfort, and Philip Hollom. The book was the product of Anglo-American cooperation and drew on innovations in American field guides produced in the first third of the twentieth century.

Roger Tory Peterson, the American bird artist, had been part of this earlier period of innovation and had put together an influential field guide to the birds of the eastern United States in 1934. His approach drew strongly upon the pioneering identification work of Ludlow Griscom, and it sought, in Tory Peterson's words, 'to furnish field marks by which live birds ... may be run down by impressions, patterns and distinctive marks, rather than by anatomical differences and measurements'. Tory Peterson stripped his bird paintings of unnecessary detail, representing only those characters that were peculiar to a species 'as compared to its near relatives'.[93] For its critics, these paintings were so schematic that they compared them to wooden bird decoys.[94]

A meeting between Tory Peterson and the British ornithologist Guy Mountfort on Hawk Mountain in Pennsylvania in 1949, both of whom were there to watch migrating birds of prey, triggered the idea for a new field guide to the birds of Britain and Europe along the lines of Tory Peterson's American book. Philip Hollom, editor of the journal *British Birds* and another pioneer of the new ornithology, was brought in as the third author. Collins in London published the book in 1954 with an introduction by Julian Huxley.[95] The internationalism of the book was foregrounded by Huxley in his introduction. Not only was the book the product of a transatlantic collaboration, but it also sought to promote 'international liaisons' between naturalists in Western Europe.[96] The decision to include not just birds occurring in Britain but to extend this to include those present as breeding birds or regular passage migrants in Western Europe was deliberately aimed at a market of affluent birdwatchers (like Huxley) who already travelled to Europe to watch birds and at those whom it hoped might do so in the future.[97]

A Field Guide to the Birds of Britain and Europe, like Tory Peterson's earlier books, emphasized an approach to bird identification that was built around knowledge of 'field marks'.[98] To this end, the book enjoined its readers to memorize the key feather tracts and bare parts that constituted the 'topography' of a bird. These feather tracts were referred to by the technical terms used to distinguish areas of the bird and included key features like eye-stripes, wing-bars, and rump patterns.[99] The path-breaking feature of the book stemmed from the way this approach to identification was carried into the illustrations that lavishly filled out its pages. Unlike established forms of bird painting, Tory Peterson's illustrations were not portraits of birds in naturalistic settings but were, as he put it, 'patternistic and functional', with each bird set against a plain background in a tabular arrangement. Developing the technique he had used in his American field guide, birds were drawn to scale and similar species were shown next to each other. Arrows were used to draw attention to key 'field marks' in a mixture of black and white and colour illustrations. The aim was to produce what sociologists John Law and Michael Lynch called 'a descriptive organization of seeing'. As they suggest, Tory Peterson's illustrations were 'instructive diagrams' that aimed to highlight 'authoritatively recognizable similarities and differences'.[100]

THE ETHICS OF BIRDWATCHING

A Field Guide to the Birds of Britain and Europe set the standard for field guides over the next forty years and was part of a swathe of books that emphasized field identification and the observational and recording techniques of the new ornithology.

In a series of talks and lectures about birdwatching delivered to naturalists trusts, civic societies, and RSPB members, Peter Conder, the RSPB secretary, promoted this approach to enjoying birds, often amplifying themes from his talks in editorials in *Birds.* These talks, delivered in the early to mid-1970s, registered the growing number of people being attracted to birdwatching and the need, from Conder's perspective, to reinforce how the hobby should be conducted. In a lecture on birdwatching from February 1974, Conder emphasized the importance of record keeping for those who wanted to be a 'regular bird watcher'. This started with making field descriptions of birds to facilitate their identification and should then progress to recording songs, calls, and aspects of behaviour. At each stage the new birdwatcher needed to cultivate the 'discipline to become an … exact observer', beginning with learning the feather tracts of a bird and how to look for 'prominent field marks'.[101]

Conder expanded his thinking about the practice of birdwatching in an RSPB book published in the late 1970s called *RSPB Guide to Bird*

Figure 2.2 Identification plate, Roger Tory Peterson, from *Birds of Britain and Europe* (1954).

Watching. It explicitly aimed to induct those individuals and families who had recently become interested in birds into the deeper satisfaction of bird study. As Conder put it in the book's introduction, many people had not yet 'discovered how tightly the study of birds can enthrall them'.[102] The book aimed to turn its readers into 'bird watcher–naturalists' who watched, recorded, counted, and studied birds.[103] This was a model of birdwatching that, as Conder himself acknowledged, built directly upon Max Nicholson's *The Art of Bird Watching*.[104] It also, as we will see later, drew Conder into the moral geography of birdwatching and into some of the problematic behaviour of modern birdwatchers.

Peter Conder's promotion of birdwatching as bird study reflected his own formation as a birdwatcher in the 1930s and 1940s and the influence of the new ornithology upon him. Not only had he been inspired by the new ornithology as a schoolboy – redirecting his egg collecting hobby, like others, into first counting nests and then counting and observing birds – but he had also undertaken bird study, as we saw in the last chapter, while a prisoner of war in Germany (alongside John Buxton and George Waterston). In his RSPB book he used this experience to underline the importance of note-taking to birdwatching, emphasizing how 'even when I was a prisoner of war I managed to keep records of the birds I saw, even though paper was in very short supply'.[105]

After the war Conder became warden of Skokholm Bird Observatory (between 1947 and 1953). Skokholm, like other bird observatories, enjoyed close links with the BTO through its Bird Observatories Committee, and Conder pursued life-history studies and extensive bird ringing while acting as warden.[106]

In 1960 he was elected to the council of the BTO and repeatedly encouraged RSPB members to join the BTO and support its studies.[107] Conder was not alone among those at the upper echelons of the RSPB in having these strong ties with the new ornithology. The chairman of the RSPB council at the start of Conder's tenure as the Society's secretary was the Reverend P.H. T Hartley. Hartley had conducted research through the Edward Grey Institute, which was affiliated to the BTO, in the 1930s, and he published articles in the BTO's new journal *Bird Study* in the 1950s. Philip Brown, Conder's predecessor as RSPB secretary and head of the RSPB's Watchers and Sanctuaries from the late 1940s, served as honorary treasurer on the BTO council and was a part of the Trust's BBC Advisory Panel.[108] As

RSPB secretary from 1952 to 1963 he was a key figure in the Society's immediate post-war development. And Conder, Hartley, and Brown were not the only figures who linked the RSPB and the BTO during this period. There was also a limited crossover in membership between the two organizations, with 6 per cent of the RSPB's membership in 1968 also being members of the BTO.[109]

This convergence of approach to birdwatching within the RSPB and the BTO led some BTO insiders to see the conversion of the RSPB to the virtues of 'field study' as one of the Trust's greatest successes.[110] The idea that enjoying birds should be directed and have some purpose – be that a better understanding of your local birds or participating in a national sample census – was a message emphasized by Peter Conder in his lectures and writings, and during his leadership of the RSPB. This focus was partly driven, as I noted earlier, by the surge in birdwatching as a popular recreation and partly by a perception of the hobby's changing social profile. Conder was keen not only to instil the principles of birdwatching as bird study into the new cohort of RSPB members, but also to warn them about the dangers of bad birdwatching. In an article in *Bird Notes* he encouraged RSPB members to join the BTO and participate in its major new census of common birds. He presented this as a meaningful and worthwhile approach to birdwatching and as morally superior to the expanded hobby of 'rarity chasing'. This latter pursuit – 'the chasing of rarities at sewage farms, gravel pits and on the coast' – risked relegating purposeful bird study to a more minor role within the culture of birdwatching.[111]

Conder's concerns were shaped by the growing mobility of an expanding birdwatching public and by books that provided information about where to find and watch birds. John Gooders's *Where to Watch Birds*, first published in 1967 and running to many editions, became the focus of concern for those at the upper echelons of the RSPB about the problem of disturbance to birds caused by growing numbers of 'bird-spotters and listers'. In a review of Gooders's book in *Birds*, David Lea, the RSPB's reserves manager, pondered whether the book should have been published at all. Established birdwatchers, he confessed, often wanted to protect their favourite areas and 'feel that any publicity will bring hordes of people to drive away the birds they have come to enjoy'.[112] Alongside the problem of the unwitting disturbance caused by the growing number of mobile birdwatchers, Lea railed against the more sinister threat posed by 'ruthless' tick hunters, who stopped at nothing to get their bird.[113]

Peter Conder was also critical of Gooders's book in his *RSPB Guide to Bird Watching.* He compared it negatively with Max Nicholson's *The Art of Bird Watching,* suggesting that Gooders's book concentrated not on bird study but on '"bird listing" or "bird spotting" [and on] telling the reader the best places to go to see the greatest number of birds'. Conceding that all birdwatchers 'bird listed' to some extent, he cautioned the new birdwatchers about the risks of being lured into 'listing' as the only kind of birdwatching. As he put it, 'there is a tendency, common amongst many who have taken up bird watching recently, to think that listing birds is the only thing that matters or can be undertaken by the average bird watcher'.[114]

Conder returned to this theme in a lecture on 'Present day standards of bird watching' delivered to the Lincolnshire Naturalists Union. Picking up on an incident involving the disturbance of birds and damage to property caused by 'rarity chasers' on the Scilly Isles in 1971, he railed against this growing problem. A lot of birdwatchers, he suggested, seemed to go out looking for birds and did not know what to do when they found them: 'Their birding consists of going to well-known birding spots, ticking off lists in an increasingly sporting and competitive way... They are selfish ... because of their disregard for the bird's welfare, other birdwatchers and landowners. This activity has a variety of names such as "rubber necking", "tally-hunting", "list ticking" and now it is known as "twitching".'[115]

The antidote to this growing problem was, for Conder, the education of birdwatchers about the value of bird behaviour study or their induction into thinking 'ecologically' about birds. His aspiration was to turn 'tally hunters' into 'curious naturalists'. The latter was a term taken by Conder from ethnologist Niko Tinbergen's book on his formation as a naturalist–scientist.

Others within the RSPB, the BTO, and the journal *British Birds* shared Conder's fears about the growing problem of 'bad bird watching', and in February 1971 they joined forces to produce a statement. It was summarized in *Birds* under the subheading 'Be considerate' and noted the increasing problem of disruptive behaviour among birdwatchers in general and particularly among the 'hordes of visitors' who were attracted by rarities. 'Some people,' it chided, 'are giving the rest of us a bad name... [They] seem to be unnecessarily selfish towards other bird watchers and towards the birds themselves.'[116]

The rarities committee of *British Birds* had coordinated the joint statement. The committee had been established in 1958 to manage the huge increase in sight records of rare birds that were being submitted to the journal. It aimed to support county bird recorders concerning the criteria for accepting these records.[117] This was to ensure that the true status of rare birds was correctly recorded in the county bird reports and archived within the journal in scientifically robust ways.[118]

A later editorial in *British Birds* reaffirmed the work of the committee, but this time it offered a more sympathetic view of 'rarity chasing' – or what it called the 'enjoyable sport of spotting rare birds'. The editorial argued that through the work of the rarities committee, interesting patterns had been noted for individual (rare) species that justified the attention focused on rare birds.[119] The chair of the committee, D.I. M Wallace, was even more robust in his defence of rarity chasing. Reflecting on the first ten years of the rarities committee, he noted the increase in field identification skills since the early 1950s and the upsurge in observations. This meant that more rare birds were being found. 'The rarity,' he suggested, 'always the prize for ordinary bird watchers, [had] passed from being an occasional delight to a constant goal for some and a more common by-product for others.' For Wallace this was a good thing as it provided data for analysis on a national scale and a fuller sense of Britain's avifauna.[120] In a later – breathless – review of the host of rare birds seen on the Isles of Scilly in the autumn of 1971, he again asserted that the 'enjoyable sport' of rarity hunting could be combined with science.[121]

Wallace's comments were notable for recognizing and affirming the sporting element of rarity chasing. He was not unique in doing so. There were also celebrated precedents within American birdwatching. In his introduction to the *Field Guide to the Birds of Britain and Europe*, Roger Tory Peterson had acknowledged that 'it is the discovery of rarities that puts real zest into the sport of birding, a zest that many of us would like to interpret as "scientific zeal" rather than the quickening of our sporting blood'.[122] Tory Peterson's investment in and acknowledgement of the sporting and collecting aspects of 'listing' revealed, like Wallace's comments, the strong parallels between elements of the expanded culture of birdwatching and older bird-centred cultures of nature. The 'zeal' for seeing rare and scarce birds and building a 'collection' of life ticks strongly echoed the excitement and adventure of bird's-nesting and the impulse to generate a collection of eggs. The 'lists collectors' of the mid–late twentieth century shared a commitment to the adventure of the quest and the value of the rare and unusual with the collectors of birds' eggs.

CONCLUSION

In an editorial commentary from the spring of 1972 in *Birds*, Peter Conder had ventured a partial revision of his promotion of purposeful birdwatching in a comment piece to RSPB members. His column sought to legitimate what *British Birds* (BB) had described in its own editorial a few months earlier as 'lazy' and 'selfish' birdwatching. While BB principally had in mind those active birdwatchers, including 'rarity chasers', who 'simply watch birds for pleasure and not for posterity', Conder took the criticism to include the majority of the RSPB's membership who were casual birdwatchers and who had never considered submitting bird records to either county or national recorders. In reflective mood, Conder defended these 'ordinary', casual birdwatchers, an important part of the RSPB's growing membership, who simply took pleasure and joy from seeing birds. They formed a growing part of the majority of RSPB members who owned garden bird feeders and who had at least one nest box in their gardens.[123] Their emotional relationship with birds, Conder argued, was an essential element of bird conservation: a bedrock of feeling that underpinned the support for the welfare of wild birds and bird protection within society at large. As he put it: 'Those whose pleasures are simpler and who enjoy seeing birds for their own sake without wishing to analyze, to quantify or to record are doing something equally important. When it comes to the test, their affection and concern for birds is the strongest weapon we have.'[124]

This was a significant softening of the vision for purposeful and collective birdwatching that he had promoted and would continue to promote in his talks and writings and in his leadership of the RSPB. His defence of casual birdwatching and the debate around the 'lazy' and 'selfish' birdwatcher revealed the divisions between different groupings within the expanded hobby of post-war birdwatching as well as the ethical or moral judgements that cut across this culture. It was a debate that generated its own moral geographies of birdwatching and of how birdwatchers should conduct themselves in their encounters with wild birds and the urban and rural spaces in which they were found. For the defenders of bird study and the ethic of the new ornithology, casual bird lovers were less of a problem than obsessive and destructive 'rarity chasers' or 'tally hunters'. We have seen how these figures were viewed within the RSPB, in particular, as 'bad birdwatchers' against which legitimate ways of enjoying birds were defined.

The model birdwatcher of the immediate post-war decades was defined as a rigorous observer and recorder of birds, attuned in particular to the demands and pleasures of bird study. This model, institutionalized within the RSPB from the late 1940s and evident in the county bird societies and the BTO's ongoing studies, achieved much of its force in the campaign led by the RSPB to educate children and their parents and teachers about the 'life destroying' hobby of bird's-nesting. This chapter has sought to recover the central role that bird's-nesting played within post-war children's culture and the legacy of its influence upon earlier generations. The established status of this hobby and its legitimacy required not only a sustained campaign of law reform and enforcement on the part of the RSPB, but also an educational and propaganda campaign to convert bird's-nesters into birdwatchers. Central to this work was the suppression of attention to birds' eggs within popular natural history and the attempt to exclude them from the birdwatchers' field of vision. The RSPB had some success in this strategy, and birds' eggs were marginalized within the culture of birdwatching by the late 1970s, disappearing from bird books and not looked for by the great majority of birdwatchers. The 'schoolboy bird's-nester' that parliamentarians in the 1950s had sought to protect quietly slipped away, surviving only in the guilty recollections of committed birdwatchers and as a symbol of an older, more destructive bird-centred culture of nature. Legal reform and the RSPB's propaganda, however, did not manage to totally eradicate the practice of egg collecting, and a small number of adult collectors remained a persistent problem for conservationists and the police. By the late 1970s, however, this residual culture of egg collecting was firmly located within the sphere of wildlife crime and had been stripped of any legitimacy as a nature culture.

As I have argued in this chapter, the collecting and sporting impulse that defined bird's-nesting did not disappear from human/avian relations, however, but instead found new expression among the post-war birdwatchers in the collecting of lists of birds and in the adventure of rarity chasing. Bird listers shared with egg collectors the pleasures of the chase and the emotional pull of adding to one's collection – features common to collecting cultures in general.[125] Bird listers and egg collectors further located their prized objects within what cultural critic Walter Benjamin called the 'magical space' of the collection. In the case of egg collectors this took the material form of boxes, drawers, and cabinets. For bird listers, records of birds were held in notebooks and remembered and recounted to a community of fellow birdwatchers. Both forms of collecting, however, were shaped by similar desires: the desire to understand, to wonder at, to control, and to order the 'apparent infinitude of animate and inanimate things'.[126]

The competitive collecting impulse was strongest among 'tally-listers'. For this subset of birdwatchers, the 'occasional delight' of the rarity became the central objective of bird spotting. Facilitated by growing affluence and mobility, the observational culture of this kind of birdwatching was geared not towards 'bird study' but to seeking out the rare and scarce. Recording birds was stripped down to the day list, the year list, and the life list. Part of the allure of rare birds may have been, as Richard Mabey has suggested, the sense of privilege that these observations conferred upon the observer. As in other fields, including within other forms of collecting, rarity was the source of value and enhanced the symbolic status of the desired object. Seeing the rare bird and adding it to your list allowed the bird spotter to appropriate that value in the form of kudos or status among their peers. This was birdwatching understood as a tournament of value.

Perhaps one of the paradoxes of the contestation over the ethics of birdwatching during the immediate post-war decades was the way the RSPB faced in both directions on the issue of rare birds. While it championed an attention to common and abundant birds within the promotion of collective bird study – particularly under the leadership of both Philip Brown and Peter Conder – much of the Society's protectionist and conservationist work focused on rare birds. As publicity for the BTO in the late 1960s put it, 'the rare birds are being … looked after by the RSPB … the BTO are also very concerned about the common ones'.[127] The attention given to rare birds within much of the RSPB's post-war protection activity could reinforce the allure of the rare and scarce in ways that appeared to cut across its broader mission to encourage parochial observation among its membership. Rare birds, particularly charismatic species like the osprey and avocet, mattered more and more to the RSPB because of their appeal to a large public of bird lovers. These rare birds became central to its publicity and campaigning and were woven into the modernizing of the RSPB's approach to what it came to increasingly call bird conservation. As we will see in the next chapter, bringing the public into close encounters with these rare birds formed a large part of the RSPB's management of its bird reserves.

3

'A Close-up View of Birds': Nature Reserves, Conservation, and Watching Birds

In May 1963 the Council for Nature, a body representing more than 300 wildlife and countryside groups, organized the first National Nature Week (NNW). Its aim was to mobilize public opinion in support of the work being done by the voluntary conservation movement and to communicate the new and urgent pressures on the British countryside and its wildlife.[1] To support NNW the Post Office issued two commemorative stamps and the *Observer* journalist Richard Fitter published a guide to the countryside titled (with more than a nod to Cobbett) *Fitter's Rural Rides.*[2] A series of nature trails were also opened across the country, bringing this American innovation in countryside recreation to the United Kingdom for the first time.[3] At the heart of NNW, however, was an exhibition sponsored by the *Observer* and held at the Royal Horticultural Halls in Westminster. The exhibition included displays from the World Wildlife Fund, the National Trust, Nature Conservancy, the Wildfowl Trust, the Botanical Society of the British Isles, and the Royal Society for the Protection of Birds (RSPB). Across the week, 46,000 people visited the exhibition.

It is the RSPB's display that particularly interests me here. At its centre was a reproduction of the view from the interior of an observation hide. This simulated the perspective of visitors to the RSPB's three most important nature reserves – Minsmere, Havergate Island, and Loch Garten – as if viewed through the horizontal opening of a hide. Back-projected films of the birds that were 'characteristic' of each reserve – the avocet at Havergate Island; the osprey at Loch Garten; and the bittern, marsh harrier, and bearded tit at Minsmere – ran on a loop under the dioramas.[4]

It was telling that the RSPB should place observation hides symbolically and materially at the centre of its stand at the NNW exhibition. The Society had constructed its first observation hides, or 'observation huts', at Minsmere in the late 1940s, but they remained a relatively novel built form.[5] By placing a simulation of one at the heart of its display, the RSPB was registering the unfamiliarity of the general public with hides, but it was also signaling their importance to the Society in allowing it to offer visitors to its reserves 'superb close up views of birds', as the exhibition's accompanying text put it.[6]

The construction and placing of observation hides on RSPB reserves formed part of a broader approach to bird conservation developed by the Society from the late 1940s. Owning or leasing sanctuaries or reserves – tracts of private land set aside for the protection of typically rare or scarce breeding or migratory birds – was central to this approach. The RSPB had first become interested in establishing 'sanctuaries' in the late 1920s. The policy had been driven by a perception of the rapid changes unfolding in the British countryside from ribbon development, tourism, and industry in the inter-war years. Experience of wildlife protection in the Dominions of Empire, where national parks and wildlife reserves had been established from the last decades of the nineteenth century, was an additional, important spur. The RSPB's interest in sanctuaries, however, was given a new focus and direction from the early 1940s by debates about the place of nature conservation in post-war reconstruction. The war years and those immediately after it, in fact, were a period of new thinking about the role of nature reserves: a period in which competing conceptions of reserves jostled for influence and in which there was a broader reconfiguration of approaches to protection, preservation, and conservation.

This chapter reflects on the role played by the RSPB in this wider debate between voluntary conservation organizations and State policymakers about the educational, promotional, and scientific role of nature reserves in the wider cause of wildlife conservation. I focus, in particular, on the reconfiguration of nature reserves and bird sanctuaries as modern spaces of encounter with wild birds. The designed environment was central to this process. Observation hides and screens, together with interpretative signage and trails, formed 'instructive landscapes' that linked people and birds to a managed environment.

This chapter focuses on the shaping of the behaviour of both birds and people at the Society's Minsmere reserve on the Suffolk coast, one of those showcased at NNW. The approach developed by the RSPB at Minsmere formed part of a new approach to bird protection that was pieced together by the Society across a number of reserves from the late 1940s. It was an approach rooted in a shift from an exclusive emphasis on secrecy and protection to one centred on controlled access and the orchestration of human–avian encounters. It was also an approach committed to the active management and manipulation of habitats rather than simply the passive preservation of existing landscapes. This new approach to the management of birds, people, and land use did not, however, represent an absolute break from the Society's inter-war view of bird protection.[7]

The RSPB continued to focus on the need to defend birds from a range of perceived threats: egg collectors most importantly, but also 'illegal' game keeping, irresponsible birdwatchers, and the 'rubber-necking' general public. Managed access and the orchestration of attention, as we will see, still involved high levels of security enacted through barbed wire, keep-out signs, and (occasionally) armed watchers and the tight control of information about rare birds. But these techniques now increasingly sat alongside the opening up of human–avian encounters.

This shift in the RSPB's approach to protection was part of a wider modernization of the Society from the early 1950s in which it developed a more assertive approach to bird conservation, campaigning, and publicity. Acquiring land for reserves was central to this process. Not only were reserves important in supporting the protection of rare birds, but places like Minsmere, Loch Garten, and Havergate Island also became important 'shop widows' – as the Society's secretary Peter Conder put it – for publicizing the RSPB's activities and growing its membership.[8] Through this, the Society believed, it could increase its income and in turn its political influence on policymakers.

In developing these arguments, the chapter draws on and seeks to extend recent environmental histories and cultural geographies of nature conservation and countryside recreation. In particular, the chapter picks up on David Matless's insistence on attending to the models of human conduct forged within nature conservation alongside, as he put it, 'the animal, vegetable and mineral'.[9] For Matless, nature conservation, especially from the 1940s, promoted an ethical vision of natural history in which the 'reserving' of nature worked alongside the promotion of reserved forms of human conduct.[10] He has also argued that the adoption of nature trails in the United Kingdom through the initiatives led by the Council for Nature can be understood as 'instructive landscapes' that aimed to educate the public about the natural environment and to shape how they conducted themselves in relation to it. One strand of this, Matless and his co-authors contend, was the ambition to normalize the appreciation of nature and to remove its association with eccentricity.[11] To this end, nature trails used marked paths, signage, and maps to encourage an attentive and ordered encounter with nature, be that urban or (more likely) rural.

The Council for Nature, however, was clearly not the first, nor the only, conservation body to develop the educational and promotional dimensions of conservation landscapes. In its approach to nature reserve management, the RSPB promoted a moral geography of leisure that aimed to shape the conduct of its visitors and to segregate 'responsible birdwatchers' from other countryside activities. In doing so, the Society used designed objects, built forms, texts, and manipulated landscapes to form an assemblage or set of devices that made possible certain kinds of bodily and visual agency in relation to wild birds and nature landscapes. The hides, screens, signage, and footpaths used on nature reserves – together with the technology of binoculars, telescopes, and cameras – organized and extended human vision and facilitated new ways of seeing and new forms of embodiment. Together with cultural forms like field guides, countryside books, magazines, and documentary films and television they constituted a regime of looking at wild birds and a way of styling the human in relation to the avian.

Understanding how the assemblage of material, technical, and textual elements shaping the 'interpretative landscapes' of bird reserves came together requires a broader exploration of the wartime and post-war conception of nature reserves in Britain. In the chapter's first section I look at the nature reserves movement and its influence upon the 'reserving of nature'. In the second section I narrow my focus and explore the work of the RSPB and its management of birds and people at its flagship reserve, Minsmere. And in the third section I consider the complexities of the observational culture orchestrated in the 'contact zones' between birds and people on the RSPB's reserves, and I consider how the aim to prescribe ways of looking could be cut across and subverted by competing ways of attending to birds.

SANCTUARIES AND NATURE RESERVES

In early 1930 the RSPB member's magazine published a lengthy article setting out the Society's intention to secure an area of Romney Marsh in Kent as a 'sanctuary'.[12] This move to 'safeguard and secure' nature through the acquisition and reserving of land had been prompted by a recent international conference on the protection of migratory wildfowl held at the Foreign Office. The conference had recommended the establishment of wildfowl sanctuaries in all participating countries. The RSPB's efforts to acquire land on Romney Marsh was the 'first attempt', as the editorial suggested, to act upon the conference's recommendation and to protect the marsh and its birds from those twin threats to coastal land in inter-war Britain: 'development and bungalows'.[13]

This concern with the 'reserving of nature' within bird sanctuaries followed the earlier, parallel, moves made by voluntary organizations like the National Trust to hold land 'in trust' in order to protect places of natural beauty and buildings and monuments of historical value. It was also stimulated by the emerging debate about the question of national parks and the perceived poverty of Britain in this regard, certainly when compared with the protection of wildlife within national parks in the United States and the Dominions of Empire and some European states (notably Holland).

In seeking to acquire sanctuaries as a key instrument in its policy of bird protection, the RSPB was breaking new ground as an organization. While it continued to campaign for legislation to limit the plumage trade and to seek greater protection for seabirds from hunting, egg collecting, and oil pollution, the Society shifted its focus to the protection of rare and scarce breeding birds in the United Kingdom. This move was directed towards what the RSPB saw as the proper implementation of the United Kingdom's myriad of local bird-protection orders. Since 1905 it had supported this through the creation of a network of 'watchers' to protect rare breeding birds. Overseen by the Watchers' Committee, 'watchers' were typically local men, often retired policemen or servicemen, employed to watch over sensitive sites and deter egg collectors and the curious public.[14] While watchers remained central to the RSPB's approach to bird protection, the acquisition of sanctuaries or reserves offered a more permanent solution to the problem of bird preservation, with the land and its birds safeguarded 'not for today only but for future generations ... vested in the Society permanently and unchangeably'.[15]

In linking the acquisition of land with the protection of rare breeding birds, an editorial in *Bird Notes and News* set out a conception of bird protection that would run through the Society's work for much of the next fifty years.[16] The RSPB's approach to sanctuaries was developed through the active role that the Society – and especially Geoffrey Dent, the chairman of its Watcher's Committee – played in wartime and early post-war debates about the role of nature reserves in the conservation of British wildlife. Dent was a director of the brewery Truman's, 'part old school preservationist, part active conservationist', who combined enthusiasms for field sports, field naturalism, and conservation.[17] He was a driving force in the moves to place the question of nature reserves at the heart of wartime debates about national parks. The Society for the Protection of Nature Reserves (SPNR), formed in 1912, had played an important role in the early years of the century in developing the idea of nature reserves in Britain and in seeking to establish a list of sites that should be preserved in this way. The SPNR had failed, however, to sustain the campaign for nature reserves, especially after World War I.[18] It was Dent who – along with other leading figures within natural history, such as Arthur Tansley – was instrumental in rekindling the promotion of nature reserves and in welding together a nature reserves movement in the 1940s. It was this movement that ensured that wildlife conservation was taken seriously within the plans for protecting the countryside within post-war reconstruction.

As early as March 1940 Dent was developing his idea of the role of sanctuaries. In discussions about establishing a watcher in the Spey Valley in the Scottish Highlands, Dent urged the RSPB's secretary R. Preston Donaldson to press 'for sanctuaries for certain species as part of any government bird protection policy after the war'.[19] Dent defined these as 'breeding sanctuaries' and suggested that a range of 'types of terrain' that supported rare and scarce breeding birds should be compiled. The terrain included heathland, marshes, and mountain areas, and each area was to be linked to specific breeding birds. As Dent noted to Donaldson, 'even two or three real sanctuaries of each type of terrain would be enough for a start... I know this is a bad time, but the groundwork should be laid in preparation for the future or we shall lose several more breeding species in the next 20 years.'[20]

A year later, in 1941, Dent worked up his ideas into a fuller memorandum that was shared with the SPNR. The document became the basis for the deliberations of a conference and, later, a government-sponsored committee – the Nature Reserves Investigation Committee (NRIC) – on the role of wildlife protection in any future national parks legislation. The NRIC's work was significant because John Dower, the leading architect of national parks, had developed a vision that focused primarily on a landscape aesthetic and on the role of national parks as places of recreation and public enjoyment (the themes of 'access' and 'amenity'). This neglected the wildlife value of the parks – something that the NRIC sought to rectify.

Following the work of its regional subcommittees, the NRIC published a list of proposed national nature reserves (NNRS) in 1945. These had been selected principally for their ecological value, as 'representative samples of the British flora and fauna'.[21] The NRIC's memo 'Nature conservation in Britain', published two years earlier in 1943, elaborated a model of how conservation should work in these national nature reserves. It drew a threefold distinction between conservation as (1) the pursuit of scientific and economic studies, (2) the enjoyment of nature by the public, and (3) the 'promotion of education in natural history'.[22] This definition, especially its first element, evidenced the growing influence of the relatively new science of ecology and of ecologists on the thinking about the management of natural landscapes. As Stephen Bocking has shown, ecologists like Arthur Tansley, who had coined the concept of the ecosystem, promoted the active management of nature reserves rather than simply the 'preservation' of habitats and rare species. The role of conservation was to actively manage these habitats contained within nature reserves in order to maintain their 'distinctive plant and animal communities'.[23]

When the government legislated in 1949 to establish national and local nature reserves within the terms of the National Parks and Access to the Countryside legislation and to create the Nature Conservancy, it drew on the NRIC's recommendations – particularly about the role of nature reserves as sites of scientific value.[24]

The Nature Conservancy, charged with overseeing and designating national nature reserves, adopted a view that the most important role of nature reserves was as 'open-air laboratories' from which the public was excluded. Max Nicholson, the director of the Nature Conservancy from 1952, was forceful in driving forward the establishment of national nature reserves, with seventy being created by the early 1960s during his tenure at the conservancy.[25] In supporting the creation of national nature reserves, Nicholson consistently pressed for a planned and systematic approach to their selection and the idea of reserves as primarily 'living museums and outdoor laboratories'.[26] The aim of nature reserves, in Nicholson's view, continued to be the safeguarding of a 'fully representative selection of habitats allowing the systematic study of their ecosystem, biological productivity, population dynamics, succession and land use history'. Management of these reserves was to be 'systematic and scientific', rather than being 'achieved empirically'.[27]

Within the wider voluntary conservation sector, however, a different conception of nature reserves was articulated. Alongside his ideas of 'species sanctuaries' and 'breeding reservoirs', Geoffrey Dent suggested that the promotional dimensions of nature reserves should be exploited. In a note from August 1942 he described 'amenity reserves' that would give the 'general public the opportunity of seeing and learning to appreciate the beauty of wildflowers, trees, butterflies and birds'. Elsewhere he contended that without some 'good propaganda reserves', public support for the creation of nature reserves would not be forthcoming.[28] This idea, taken up by others within the RSPB through the 1950s and 1960s, was central to the transformation of the Society's approach to bird conservation and how it aimed to manage both birds and people on its reserves. Minsmere, an area whose potential Dent was among the first to see, became a key reserve for the Society in pursuing this new approach to 'birds and people'.[29]

MINSMERE: FROM SECRECY TO MANAGED ACCESS

Although the RSPB, and Dent in particular, had been looking at the potential of Minsmere Levels as a reserve for some time, it was the arrival and nesting of a group of avocets, a rare wading bird, in the spring of 1947 that catapulted the area to the top of the list of priorities for the Society.[30] Avocets had been extinct as a breeding bird in Britain since the 1840s, and the discovery of four pairs on nests on the Minsmere Levels on the Suffolk coast was a major event for the Society.

While the area would become a testing ground for Dent's dual ideas of a 'species reserve' and a 'propaganda' one, the RSPB's initial response was to implement its established approach to bird protection. This involved mobilizing the support of a local group of watchers to guard the birds and to impose a tight cordon of secrecy around them. Such was the rigorous imposition of this secrecy that even within the RSPB most staff did not know about the Minsmere birds.[31]

When a second group of avocet colonists were found that same spring just a few miles down the Suffolk coast from Minsmere at Havergate Island, only Dent, as chairman of the Watchers' Committee, knew of both groups of birds.[32] Thus, in 1947, two separate operations to safeguard and protect groups of Suffolk avocets were established without either group of human watchers being aware of the other's existence.

For Dent, and the RSPB hierarchy in general, there was good reason for this high degree of secrecy. They saw the threats facing rare breeding birds like the avocet as high, coming mainly from egg collectors and from unintended disturbance by curious members of the public. For those involved in the protection, these threats were repeatedly coded through a set of militarized metaphors. The birds, for their protectors, had defined enemies, especially egg collectors and the curious, but also natural ones like rats and foxes. Colonel J.K. Stanford, one of the voluntary watchers on the Suffolk coast, published an account of the discovery of avocets at Minsmere in 1947 and the subsequent watching of Havergate Island that was shot through with military protocols and metaphors.[33] Stanford used the convention of wartime code words and the drawing of an exclusionary cordon of secrecy around the rare birds to segregate them from the public and to reinforce the privileged status of the minority of trusted watchers who knew about these secret places. Stanford and others referred to Havergate Island, where the black and white avocets were breeding, as 'Zebra Island' to protect its identity even within the RSPB. At Minsmere, Stanford noted, the landowners had to be bound by secrecy and everyone had to 'lie like troopers'.[34] In reflecting on the success of the watchers at protecting the birds on Havergate Island, Stanford concluded that 'campaigns in aid of rare birds are won by victories over vermin and egg snatchers of every kind from rats upwards'.[35]

This militarized understanding was undoubtedly encouraged by the proximity of the war, and by the fact that Minsmere Levels had been a battle-training area during the hostilities and was not de-requisitioned by the army until two years after the RSPB began watching the area. The landscape of the levels was also still littered with evidence of wartime defences and mortar bomb holes. As Ben Garlick and Sophia Davis have shown, this militarization of bird protection was deeply rooted within the RSPB through the late 1940s and 1950s, being central to how key protagonists in the Society undertook bird protection at Loch Garten as well as at the Suffolk reserves.

For Garlick, the militarization of bird protection was a product of a post-conflict society. One dimension of this concerned the way the RSPB's post-war work was led by a group of men who brought their wartime experiences to bear on the practices and discourses of conservation. The RSPB secretary Philip Brown and his successor Peter Conder – who were both personally involved in the protection of ospreys at Loch Garten and of the Suffolk avocets – had served in the war. Herbert Axell, Minsmere's warden from 1959 and also involved in the guarding of the Loch Garten ospreys, had served in the Royal Artillery, seeing action in North Africa and Germany. George Waterston, the architect of 'Operation Osprey', the most militarized of the Society's bird protection schemes, had also seen active service and had spent time with Peter Conder as a prisoner of war in Germany.[36] As Garlick strikingly puts it, these men (specifically Conder and Waterston) brought 'a certain "POW mentality" to bird protection … via the the camp like materiality of surveillance … barriers and barbed wire'.[37]

For Sophia Davis, the militarized nature of conservation on the Suffolk coast stemmed from the way the secrecy and privacy of this landscape was extended from its military to its conservation use. Like battle training areas, nature reserves (Minsmere and Havergate Island) were initially shaped as 'isolated, separate places'.[38]

For all its tight surveillance of rare birds and the secrecy surrounding them, however, the RSPB's 'militarized' approach to protection could not always safeguard birds from human and animal 'enemies' and 'natural disasters'. Between 1955 and 1958 the breeding pair of ospreys in the Spey Valley was disturbed or had their nest attacked and robbed by collectors on three occasions, despite the presence of armed watchers.[39] At Havergate Island rats took a heavy toll on the avocets, and a flood devastated the island in 1953. At Minsmere the problem was the birds themselves – or rather the habitat. After breeding successfully in 1947, avocets failed to return to breed between 1948 and 1963 as the reserve rapidly became degraded through the drying out of its pools. In a July 1952 letter to the landowner Captain Stuart Ogilvie, Philip Brown noted the problem with the disappearance of the pools inside the seawall and that, as a consequence, 'birds ... are not really accessible to visitors'.[40] A memo from the Watchers' Committee on the management of RSPB reserves produced by Geoffrey Dent in 1956 revealed more systematic evidence of the scale of the deterioration of the reserve. As he wearily conceded, 'considerable work is required and will be required to keep Minsmere in the conditions in which we originally found it'.[41]

To this end, the RSPB appointed a new warden called Herbert Axell in 1959, who was transferred from the Society's Dungeness reserve. Axell's first warden's report for the 1959 breeding season confirmed the scale of the problems at Minsmere. Alongside the drying out of the pools and meres behind the shore wall, other parts of the reserve had lost much of their ornithological value. 'The Tree and Central hides,' Axell noted, were 'in the opinion of very many visitors, hardly worth visiting.'[42]

To address the degrading of the reserve habitat, Axell proposed creating new shallow meres by opening up 'a large area of virtually waste ground'. Central to this would be a new saltwater mere 'with bare islands for waders ... and terns'. The top layer of this 'waste ground' would be 'scraped off mechanically'.[43] Over ten winters from 1962 onwards this work was slowly undertaken, much of it by hand, and the dried-out section of the marsh was landscaped to create a series of new lagoons. The water levels on this area – which became known as 'The Scrape' – were carefully controlled, with encroaching vegetation regularly sprayed and dug out in order to prevent the natural cycle of succession occurring.

In devising the Scrape, Axell was forward looking in thinking about not just how to make the site attractive to birds, but also how to give visitors the opportunity to see the birds. In this regard he was taking up the baton from Philip Brown and, particularly, George Waterston, who had been keen to show the visiting public the RSPB's 'star' breeding birds. Brown and the RSPB council had taken the decision in 1950 to publicly name Havergate Island as the site in East Anglia where avocets were breeding and to invite a limited number of visitors with permits to view the birds from a small four-seater hide.[44] At Loch Garten, Waterston's long-held desire to make public the news of the ospreys once they had successfully bred and to establish a viewing point for tourists was implemented in the summer of 1959. Fourteen thousand people came to see the ospreys that summer, processing along a track screened by trees to a public observation hide. The birds were viewed, from some considerable distance, through binocular telescopes.[45]

But intervening in the landscape to shape the movements of birds and people was taken much further in Axell's approach to reserve management at Minsmere. Aware of the pressure on coastal sites like Minsmere beach and the land adjacent to it, Axell mixed 'access' and 'exclusion' in the management of people on the reserve. Taking the reserve as a whole, he argued that it was important to create a section where 'people could disport themselves … where some erosion of land and disturbance of wildlife would be accepted and catered for in the provision of a car park, picnic area and nature study area'. Alongside this opening up of access, however, he also advocated a policy of 'totally excluding' the public from the largest section of the reserve and 'maintaining and managing this for research and for the benefit of rare plants and animals'.[46]

Defending the boundaries of the reserve and tightly controlling the numbers of those allowed in was central to Axell's management approach. While he waited for the resources from RSPB HQ to develop the Scrape, he focused much of his energy on Minsmere beach and the seawall. This area became a fractious border in Axell's efforts to segregate non-birdwatching holidaymakers – with their cars, motorbikes, caravans, litter, and camp fires – from the tightly controlled space of the reserve.[47] In his correspondence with his bosses, Axell repeatedly fulminated about the external pressure on the reserve from the holidaymaking hordes. As he noted in June 1960, 'the people on the beach up to the first of our noticeboards was like Brighton'.[48] Axell was vigorous in his use of keep-out signs and introduced new fencing, including barbed wire, along the shoreline edge of the reserve to preserve the integrity of the RSPB's land.[49]

This policing of the boundary of the reserve as a place-apart from popular recreations could provoke hostility from those used to having access to the seawall and other areas. A local resident and RSPB member, Mr Watt, wrote to the RSPB in March 1961 to complain about the use of barbed wire and the curtailing of public access. He was astonished, he noted, 'when suddenly confronted with a barbed wire fence on the seaward side of the seawall, bearing in mind that the public has for all the years I have known and lived in the area had free access to the seawall'.[50] Invoking earlier popular resistance to restrictions placed on common land, Watt demanded a meeting with the RSPB to discuss their policy of 'enclosures'. The local Westleton Council also complained about the unsightly use of barbed wire. The RSPB replied that the wire was designed to deter not locals, whom it felt sure were sensitive to the needs of birds, but 'holidaymakers'.[51]

Axell was not troubled by these protests, and he noted with some satisfaction the success of the barbed wire fence at keeping people away from the shore wall and reducing the disturbance of birds on the reserve.[52] He was exercised, however, by the growing problem of trespass from birdwatchers, in particular the attraction of the reserve to what he called the 'tally-hunters': that is, the pursuers of rare birds. As he suggested in his 1967 annual report, 'with an annually increasing number of people on the move and the reputation of the reserve as a place for rare birds coupled with a more efficient tally-hunters grapevine, more cases of trespass were inevitable'.[53]

This trespass often took the form of birdwatchers sleeping in the public hides in order to see rare birds or entering the reserve itself without permits.[54] A party of frustrated birdwatchers from the London Natural History Society, Axell complained, had climbed through the fencing to get onto the reserve on finding that the public hide was full.[55] Axell's antipathy to tally-hunters waxed and waned somewhat during his tenure at Minsmere, and he would on occasion give them a lift on his tractor and let them see the rare birds in return for a financial donation. At other times he insisted on the need to deter them and of the importance of placing some rare birds in the 'secret category'. As he suggested, 'some secrets have to be kept from Britain's growing army of determined rarity chasers'.[56]

Egg collectors, alongside over-exuberant birdwatchers and curious holidaymakers, also remained a problem for the reserve warden. In 1972, what Axell termed a 'human predator' stole two clutches of avocet's eggs from the Scrape. To counter this threat an all-night guard was introduced, a new guard hut was built, and a powerful searchlight was deployed. Alongside lights that were kept burning overnight in the East Hide overlooking the Scrape and in the

Figure 3.1 Herbert Axell controlling reed growth at Minsmere (1960s).

warden's office 'as a deterrent', a high barbed wire gate was made to block a key point of access from the shore.[57]

If Axell's measures against the birds' various 'enemies' on the Scrape were generally effective, the wider policy of controlling access to the reserve generated additional problems for the warden and his employers. Prior to Axell's arrival at Minsmere in 1959, between 400 and 500 visitors had been allowed on the reserve between April and September, with a maximum of six people per day, four days a week.[58] Visitors had to apply in writing to RSPB HQ for a permit, with no guarantee of success. Many of those who did manage to secure a permit were, as Axell wryly conceded, drawn from Britain's ornithological elite.[59] The restrictions on access generated by the permit system frustrated those RSPB members who were excluded from the reserve. These restrictions were made worse for some when they noticed that special access was given to 'VIP' visitors. As one irate member complained, the RSPB had become an organization for 'the privileged few' who were able to enter the reserve through their social connections.[60] Axell, who was usually required to guide these VIP guests around the reserve, found that the commitment ate up his time. Apologizing for a delay in replying to the RSPB's secretary, Axell explained that he had been 'busy with visitors (Colonel JK Stanford, Lord Cranbrook, General Sir Gerald Lathbury and others not in Debretts)'.[61]

While he often complained about the number of visitors on the reserve, Axell did recognize the contradiction of seeking public support for bird conservation and the limiting of access to reserves – particularly for RSPB members. As he argued, creating disappointment among members unable to obtain permits 'was bad for our public image and would lead to more resignations of membership... From the point of view of a Society like ours, conservation should not only be done but should be seen to be done.'[62]

This view found increasing favour at the upper echelons of the RSPB. From the early 1970s the Society placed greater emphasis on their reserves as sources of publicity and new members.[63] One consequence of this was that permits at Minsmere were increased to 100 per day. During 1974, 10,000 people visited the reserve.[64] Axell himself – though at times wildly contradictory over the issue of visitor numbers – did seek to improve the experience of visitors.

His successor (from 1975), Jeremy Soreson, continued this trend, proudly noting in 1981 that more bird species had been recorded on the reserve at the same time as human visitor numbers had also reached an all-time high.[65] The inference was clear: lots of birds and lots of people could coexist if both were managed carefully. Sorenson, however, also cautioned against the dangers of over-populating the reserve. Like Axell, he believed in public access within controlled limits. Responding to a suggestion that the car park at Minsmere should be extended, he warned that increasing parking space 'would be the start of the reserve degenerating to a "Blackpool Beach" situation. We need to avoid commercializing the reserve at the expense of the natural history esthetic.'[66]

In balancing the needs of birds and people on the reserve through management, a key part of the approach of Axell and his successors to running Minsmere was to carefully structure the public's encounters with birds. To this end, the bulk of Axell's management effort went into those parts of the reserve designed to be accessed by visitors. The Scrape was at the centre of this. In Axell's vision for it, the Scrape would be a 'large, manipulated safe habitat for common and special birds, surrounded by screened tracks and big, permanent wooden hides'.[67] He called this the 'arena concept', and he described the Scrape – with its oval shape and hides on all sides – as being 'like Wembley stadium'.[68] 'We should have to accept,' he conceded, the need to become 'part of the entertainment industry.'[69]

Figure 3.2 Bird hide at Minsmere (1960s).

Careful decisions were made about what species deserved to belong on the reserve, with mammalian and avian predators and destructive species like brown rat and coypu heavily controlled. Axell also shot an unlucky fox in 1972, picked out in the spotlight that was being used to protect the Scrape from egg collectors. It was, however, the shaping of human conduct on the reserve that particularly interests me.

Axell's creation of a designed and managed landscape made extensive use of bird hides, screens, vegetation cover, marked-out trails, information boards, and, by the mid-1970s, a visitor centre. These elements worked to make Minsmere an 'instructive landscape'. The elements that made up this 'instructive landscape' sought to educate the reserve's visitors about the environment and its non-human – especially avian – lifeforms. The ambition to educate through the shaping of the landscape, however, should not be understood as simply a cognitive process. The reserve management used the design of the environment and material devices like hides, screens, and signage to shape ways of being human in the

Figure 3.3 Duck decoy at Nacton, Suffolk.

landscape. This centred on the cultivation of an attentive form of conduct that would facilitate encounters between people and wild birds through close observation. The permanent bird hides were the key socio-technical device.

As a device they drew on earlier designed objects for bringing people close to birds. One key influence was the decoy screens used by wildfowlers through the eighteenth and nineteenth centuries. Decoys combined artificially created decoy ponds and reed screens, along with decoy ducks and luring dogs. The Reverend Richard Lubbock, an observer of Norfolk wildlife in the nineteenth century, suggested in his 1834 *Sketch of the Natural History of Yarmouth* that 'perhaps the best place in which to speculate and gain knowledge on the habits of various birds [was] the view given through the reed-screen of a decoy pond'.[70] While Lubbock was happy simply to observe, most users of the decoy screen were interested in sport and killing, styling human–avian relations in a different way from that prescribed on bird reserves.

Another influence on Axell's use of hides came from bird photography. In the early part of the twentieth century, bird

Figure 3.4 Eric Hosking and gamekeeper George Boast at Staverton, Suffolk (1930).

photographers developed the idea of hiding the human form from wild birds to facilitate close encounters, creating lightweight mobile and fixed hides. Similarly, ethologists, developing the science of animal behaviour in the first part of the twentieth century, used lightweight hides to observe the natural forms of animal behaviour.

There were, in fact, increasingly strong intellectual ties between bird photographers and ethologists through the twentieth century as both focused on recording the 'private lives of birds', especially around the nest, including courtship, mating, defence and distraction display, and the care of young.

Permanent bird hides like those developed at Minsmere shared much with the observational technology of bird photographers and ethologists. But they also shaped their own ways of seeing. They offered – as the RSPB explicitly acknowledged in its display at the National Nature Week exhibition in 1963 – a 'diorama'-like presentation of an 'exemplary' selection of birds and habitats in which the human observer was present as an observing eye. At the same time, the placing of the hides brought birds up close to the observer, especially when viewed through binoculars, allowing intimate insights into the lives of wild birds and the dramas of their existence. In this sense, bird hides were material devices that at once distanced human–avian relations and facilitated the framing of a close encounter. Axell was particularly keen to organize what was visible in front of the permanent hides in order to ensure that these close encounters could happen. As he suggested, 'good opportunities for closer observation … of feeding and roosting birds are … provided by perches and islets of gravel, mud or grass made for this reason in front of most hides'.[71]

Not all of the RSPB's observation hides were as successful at staging these close-up avian encounters. At Loch Garten, despite the exhibitionary zeal of George Waterston, the public observation hide was some considerable distance from the osprey nest. Significantly, the RSPB's education officer recognized this fact when discussing the design of the RSPB's stand at the National Nature Week exhibition. He told the model-maker that 'the diorama must simulate the view from our forward hide [not the public observation point] as otherwise the scale of the actual nesting site will be too small'.[72]

The design of Minsmere's hides proved influential within the RSPB and was copied by other organizations such as the Norfolk Naturalists Trust and the Wildfowl Trust, who wrote requesting the blueprints.[73] By 1973 there were seven hides on the reserve, holding about 160 people, plus two 'free' public hides along the perimeter of the reserve on the beach. These, as Axell envisaged them, were to deter trespass and to act as 'shop windows' for the Society's work. Through the 1960s and 1970s there was an almost continuous process of building and extending the reserve hides. A nine-seater hide was built on the eastern edge of the Scrape in 1964 with a protective tunnel over the seawall and large reed screens. It was extended in 1970 to seat at least twenty people, and then extended again in 1971 and 1973.[74] A two-mile walk around the Scrape transferred visitors between the 'stands' overlooking the 'arena'.[75] Spectators were encouraged to contain their exuberance in the hides, but were 'not expected to behave with unusual quietness'.[76]

WATCHERS AND OBSERVERS

The styles of looking encouraged and facilitated by the hides worked to institute, programmatically at least, a particular observational configuration of human–avian relations. Sitting in relative comfort on a wooden bench, arms on a resting shelf, and orientated to look through the horizontal viewing slit, the human form was taken out of the landscape and the birds were brought close to the spectator.

This organization of observation was itself part of a wider regime of looking that drew on the shaping of spectatorship within cultural forms associated with the popularization of birdwatching, including the RSPB's own publications. Represented within the Society's members' magazine and in popular countryside books, this style of looking drew heavily on the precepts – if not always the precise language – of ethology. This 'ethological eye' prioritized the close visual apprehension of behaviour over sentimental or aesthetic approaches to birds.

In an essay in the RSPB's magazine *Bird Notes*, Philip Brown, chairman of the Watcher's Committee and later the Society's secretary, codified this way of looking. Reflecting on a visit to a secret site in East Anglia to observe avocets, he directed the reader to the birds' noteworthy behaviour: 'A day seldom passes,' he noted, 'without the typical gathering of a group of birds ... with much piping, grouping themselves roughly in a circle, heads bowed.'[77] Two years later, in a book produced by the RSPB, *Avocets in England*, Brown again encouraged his readers to look at avocets in behavioural terms, focusing on the birds' courtship and their communal and distraction displays. Brown's book contained precise behavioural illustrations by the renowned bird artist Charles Tunicliffe depicting a pair of avocets displaying and mating.

Two of the voluntary watchers at Havergate Island (Brown's 'secret site') took his model of observation to heart in their own watching of the birds. As the Havergate log book from 20 May 1951 records,

'Brig. Furze and Miss Gladdy saw display and coition from No. 2 hide, exactly as described and illustrated in "A in E" ("Avocets in England").'[78] An RSPB silent film *Avocet Island,* which was screened at an RSPB London meeting in the autumn of 1950 with an introduction by J.K. Stanford, similarly used sequences designed to illustrate 'the behavior of the birds, including changeover duties at the nest and quarrelling parties of adults'.[79]

The 'ethological eye' promoted by Brown was also popularized by photographers like Eric Hosking. Hosking, who enjoyed a close and privileged relationship with Minsmere, did much to create the public image of the reserve in the 1950s, 1960s, and 1970s as the home of rare and scarce birds. His photography was notable for foregrounding the 'domestic' world of birds, focusing on life around the nest. As he noted, the '[photographic] hide's peep-hole is a window on the bird's world, giving an unparalleled view of their private lives and loves'.[80] At Minsmere his photography included recording the 'avocet's family life'.[81] In his 1955 book *Fighting Birds,* co-authored with Stuart Smith, Hosking had devoted a chapter to the display behaviour of avocets. This set the terms for much of his later studies of the birds at Minsmere.

Because of the rarity of avocets in eastern England in the 1950s and the need to avoid disturbing them in the breeding season, Hosking's portraits in *Fighting Birds* had been taken in Holland, where the birds were much more common. If the British breeding avocets were, in Hosking's and Smith's words, 'sacrosanct', they were able to 'indulge ourselves … as we wished' in Holland, placing their hide within a few feet of the Dutch birds' nests.[82] Hosking focused on recording various kinds of display behaviour, including aggressive display and distraction display. In one image, a close-up of shot of an adult avocet performing a dramatic 'distraction display with wings spread and head bowed', Stuart Smith's hand is shown touching the avocet's eggs in the nest.[83] In the commentary accompanying the photographs of the birds, Hosking and Smith drew on the ethological wader studies of G.F. Makkink and his interpretation of displacement activities, including 'displacement sleeping'.[84] It showed how Hosking's photographic practice was rooted within ethological understandings of bird behaviour and its focus on the most visually striking and dramatic forms of avian action.

The RSPB's watchers and wardens were encouraged to subscribe to a rigorous version of this ethological gaze, recording in sometimes mind-numbing detail the behaviour of their star birds.[85] Not all watchers were entirely convinced of the value of this detailed observation and recording. One watcher at Loch Garten suggested that he could see little value in 'recording every time an osprey drops off the nest to circle round and crap'.[86]

Other wardens and watchers could not help projecting cultural values onto the birds, despite the injunction to observe and record scientifically. In relation to avocets, watchers repeatedly echoed a theme that had been dominant in public representations of the bird in Britain since the early part of the twentieth century. This drew attention to the aesthetic characteristics of avocets, especially their grace, beauty, and elegance. As one watcher from August 1951 gushed, the flock of avocets that flew into the lagoons at Havergate looked like a 'troupe of beautiful ballet dancers in black and white costumes, rising from behind the curtain and taking their place on the stage'.[87] A report on a guided visit to Havergate Island in the *Times* in 1953 drew the same comparison, the journalist pinpointing the avocet's 'grace and beauty' in 'balletic' performance.[88] When Havergate Island's warden, Reg Partridge, pointed out a bird performing 'distraction display' among a group of avocets, the journalist balked at this ethological interpretation and confessed that he preferred to think of it as 'ballet dancing – with a seat in a box, too'.[89] Bert Axell was drawn onto the same cultural terrain, describing the 'leggy, aristocratic elegance' of avocets at Minsmere and claiming that the hides enabled him (and visitors) to watch the birds 'as if at the ballet'.[90]

The privilege of being able to observe avocets at close quarters at a time when access for the general public remained so controlled may have prompted these theatrical comparisons. Watching avocets was, for the lucky few, like an elite cultural experience. It also speaks of the social backgrounds and cultural tastes of these observers. Julian Huxley had revealed much about his own cultural formation when he described the breeding rituals of avocets in the 1920s (in Holland) through a comparison with the dance of Macheath and Polly in John Gay's *The Beggar's Opera*.[91]

Other spectators at both Havergate Island and Minsmere invoked not so much theatrical associations in their observation of avocets from hides, as concern that the artificial environments and the managed spectatorship of birds arranged by the RSPB was too much like seeing animals in a zoo. As the bird photographer G.K. Yeats expressed it to a senior official at the RSPB as early as 1949, it seemed that the RSPB wanted to 'keep [the avocets] in a sort of privileged zoo for the next decades'.[92]

At Loch Garten, with its heavily protected ospreys, some of the visitors that turned up to view the birds in 1959 wondered at what time they were being fed.[93] This was not such a fanciful notion. During the harsh winter of 1962/3, Bert Axell put out fish to sustain birds on the reserve, especially the rare bitterns. He also captured a number of finches and caged them in order to feed them during the hard winter.[94] Other sceptics, like Colin Tubbs, assistant regional officer at Nature Conservancy, worried that the manipulation of landscapes and all the 'furniture of countryside interpretation' was eroding the sense of 'wildness' that had made places like Minsmere attractive to naturalists in the first place. As he sharply put it, 'you do not need to build bungalows to create suburbia in the countryside'.[95]

Established naturalists, whose field craft had been developed in an era of the careful stalking of wildlife, could also find bird hides restrictive. As the bird artist Eric Ennion complained, whilst hides could allow close views of often-difficult-to-see birds, a good deal could still be seen 'without limiting vision to a view-slit and missing all sorts of things happening outside'. The visiting public could also subvert these devices through the ways in which they used them. Axell complained that on finding hides full, people would climb onto the protective bank at Minsmere to view the marshes, flushing birds in the process. He also grumbled that 'young people played around the hides, visitors deliberately leaving the marked paths to look for nests or take photographs … and [pick] plants (one woman picked a complete colony of marsh orchids)'.[96]

Other cultures of nature could also disrupt the observational practices promoted on the reserve. Until the RSPB bought Minsmere in 1977, it continued to be shot over by its landlord from late August. Axell complained that the visitors he was escorting around the reserve often had shots ring out over their heads, scattering the birds they had come to see.[97] Jeremy Sorenson was also exercised about the effects of shooting on the estate. Ahead of the full-purchase of the reserve by the RSPB, he noted in his 1976 Annual Report: 'Nice to think this should be the last full season of pheasant, wildfowl and rabbit shooting. Most weeks of the season the reserve has to put up with major disturbance of the pheasant shoot … also the wetland in front of the Tree hide and lately even North Marsh and the Scrape have had to put up with duck shooters.'[98]

The RSPB itself also sought to overlay the visiting public's view of avocets with another message that ran alongside its popularization of the techniques of 'bird study' and the careful visual apprehension of birds. This involved the construction of a conservation narrative designed to colour how visitors saw the birds on its reserves. The visiting public, including RSPB members, were encouraged to see birds such as avocets (and ospreys) as rare and noteworthy species within a national space of conservation and, moreover, as a symbol of the Society's successful work of bird protection. From the early 1950s, the RSPB emphasized its role in supporting the return of the avocet as a British breeding bird.[99] In doing so, it drew on a deep reservoir of conservationist cultural memory and wish fulfillment. The former was informed by knowledge of late Victorian and Edwardian county avifaunas that had recorded the status of avocets as a former breeding bird. The hope that they might return was prompted by reports of vagrant avocets seen in eastern or southern England and by visits made by birdwatchers to their breeding grounds in Holland in the inter-war years. Through the 1920s and 1930s *Country Life*, the leading countryside publication, regularly published articles that expressed the wish that avocets might one day return as a breeding bird to Britain. As E.L. Turner speculated in 1930, 'it is the cherished aim of all the promoters of bird sanctuaries along the eastern coastal districts to attract the elusive avocet back to its former breeding areas'.[100] The RSPB magazine *Bird Notes and News* expressed a similar aspiration in explaining the justification for the purchase of an area of Romney Marsh by the Society in 1930. Referring to the area, the editorial suggested that 'in days gone by [it was] a well-known haunt of the avocet... To persuade this interesting and beautiful species to return to breed would indeed be worthwhile.'[101]

This idea of the loss and wished-for future return of the avocet formed part of a conservation narrative that sought to promote a more enlightened relationship with these birds when set against the wanton destruction and carelessness of 'our forebears'. For contributors to *Country Life*, in particular, the drainage of wetlands and the agricultural improvement of 'wasteland' since the seventeenth century, callous egg collectors and gunners, plumage hunters, and the taste of country folk for avocet eggs in the nineteenth century were all cast as the villains in a historical morality tale. This narrative also conveyed a strong sense of the loss of elements of the 'old countryside' that had supported these birds – a loss represented as a diminution of an aspect of the country's 'natural heritage'.[102] It set in motion a drive for reparation: the making good of the sins of the past through the restitution – the return – of a once-lost bird.

When avocets did return in 1947, displaced from their Dutch breeding grounds as a result of World War II, the RSPB articulated the deep psychological pull of this conservationist narrative. At the same time it astutely promoted its role in this national conservation success story and in the making good of this old injury to the British countryside and its 'native' avifauna. While the relative abundance of avocets (and ospreys) across the rest of Europe was acknowledged by the RSPB, particularly in the 1970s, what mattered was their reincorporation into the national space of British birds and the RSPB's role as their champion.

CONCLUSION

In 1970 the RSPB's council approved the avocet as the official symbol of the Society.[103] The minimalist graphic image of the bird – presented side on as if in the act of feeding and emphasizing the bird's black and white tones – helped to consolidate the avocet as one of the 'star birds' in the RSPB's conservation pantheon. The fact that the birds could only be reliably seen on RSPB reserves underlined the association between avocets and the RSPB.

Minsmere, along with Havergate Island, were the two 'avocet' sites, and through the 1970s Minsmere hosted the most successful breeding colony of avocets in Britain. A 1973 film called *A Welcome in the Mud* celebrated the RSPB's achievements, showcasing the completion of the Scrape and its role in the avocet story. Axell's success at pioneering a new form of wetland management and in modernizing the RSPB's approach to both birds and people was recognized in the film and in publicity surrounding it.[104]

His achievements also caught the attention of other conservation organizations. The Norfolk Naturalists Trust's management committee visited Minsmere in the late 1970s and adopted the techniques used by Axell to develop its reserve at Cley on the North Norfolk coast. The Norfolk Naturalists Trust's Hickling Broad reserve followed suit. Axell's expertise was also called upon to export the model of the Scrape internationally. Following the purchase of the Donana Biological Reserve in Spain's Coto Donana in 1964, with funding from the recently formed World Wildlife Fund, Axell was seconded in March 1966 to assess how the site could be improved and to devise a management plan along the lines that he had developed at Minsmere.[105] The Coto Donana was the first of a series of adaptations of the Scrape that included reserves in Hawaii and Hong Kong.

It has been a central contention of this chapter that Axell's work to frame human–avian encounters was the most notable feature of his approach to nature reserve management. In this he shared much with other modernizers within the RSPB such as Geoffrey Dent, Philip Brown, and George Waterston in seeking to facilitate a popular engagement with wild birds, especially the rare and the scarce, through the orchestration of observational encounters with them.

As the RSPB grew in size and influence during the 1960s and 1970s and as its land holdings mushroomed, the organization played a key role in shaping how an increasingly conservation-minded public came to know about and experience wild birds. The designed and managed environment of the Society's bird reserves, with their range of material and textual devices, worked to produce distinctive encounters between people and birds. In mobilizing and staging a fascination with birds in this way, the RSPB used its 'shop window' reserves as key elements in its educational and promotional activities. In doing so it helped to broaden the appeal of a recreational relationship with wild birds and an ethic of conservation that became increasingly central elements in the structuring of human–avian relations in post-war Britain and beyond.

4

Landscapes of Loss: Wild Birds and Environmental Crisis

In 1960, Nature Conservancy, the government agency responsible for conservation in the United Kingdom, commissioned the British Trust for Ornithology (BTO) to report on the status of the peregrine falcon as a breeding bird. The study had been prompted by complaints from racing pigeon enthusiasts that the falcons were taking an unacceptable toll on their birds. The *Racing Pigeon* newspaper was particularly vociferous, urging the government to remove peregrines from the legal protection guaranteed by the Protection of Birds Act (1954). The BTO began its research in the breeding season of 1961, with a further survey conducted in 1962.[1] The preliminary findings were startling. Far from being a common breeding bird, as the *Racing Pigeon* had suggested, the research found that a catastrophic decline had occurred in the occupancy of peregrine territories since the early 1950s. The decline was particularly dramatic in southern England, with 69 per cent of the total of pre-war territories deserted and only two breeding pairs.[2]

In his report, the BTO's lead researcher Derek Ratcliffe speculated on the likely causes of the decline. Dismissing a decrease in food supply, disease, or direct persecution as possible explanations, he suggested that the most likely cause was that peregrines were the 'secondary victims of agricultural toxic chemicals through repeatedly taking prey which carried sub-lethal doses and so building up poison in the body'. Ratcliffe showed a strong correlation in time and space between the pattern of the decline of peregrines and the application of toxic pesticides across Britain's main arable and fruit-growing areas.

While Ratcliffe conceded that the link between the peregrine's decline and the use of toxic chemicals was largely circumstantial, he did cite one crucial piece of causal evidence. It concerned the analysis of a peregrine's eggshell broken by one of the parent birds from an eyrie in Perthshire. Analysis of the shell fragments found that they contained significant traces of p,p'-DDE, a metabolite of the insecticide dichlorodiphenyltrichloroethane (DDT), and smaller traces of dieldrin and heptachlor.[3] These were chemicals used in the spraying of crops or as seed dressings. Their damaging effects on wildlife had begun to be suspected in Britain from the early 1950s. Research on the late spraying of crops at ICI's Game Research Station in 1952 found high numbers of animal deaths.[4] Seven years later, in 1959, reports of an unidentified illness affecting foxes was detected and 1,300 were found dead, with many showing symptoms of poisoning. Soon after, mass fatalities of game birds and seed-eating passerines were noted at a number of landed estates and on farmland in eastern England.[5]

This emerging evidence prompted the RSPB to establish a joint committee with the BTO in the autumn of 1960 to collect records of bird deaths caused by toxic chemicals.[6] Across the seven years of the committee's existence a picture emerged of dramatic numbers of bird fatalities linked to the use of organochlorine chemicals in agriculture.[7] As the scale of bird deaths became evident, the RSPB and the BTO, together with the Game Research Association, pressed the government to restrict the use of the newer toxic chemicals in agriculture. Their lobbying was partially successful, and in July 1961 Christopher Soames, the Minister for Agriculture, Fisheries and Food, proposed a voluntary ban on the use of seed dressings on winter sown grain, while allowing their continued use for autumn sowing. DDT was excluded from the ban.[8]

The incidents of mass animal and bird poisonings in Britain and the British government's response to them was noted by an especially acute and influential observer in the United States, where the evidence of bird deaths from toxic chemicals was even more dramatic. That observer was Rachel Carson. *Silent Spring*, Carson's careful analysis of the effects of toxic chemicals on birds and other animal populations and their potential threat to human health, was published in the autumn of 1962. With its scrupulous evidencing of the harmful effects of the new post-war generation of agricultural chemicals, Carson's book crystallized the emerging concerns of conservationists on both sides of the Atlantic, presenting a startling and dystopian image of American landscapes, including suburban gardens, denuded of the colour, beauty, and song of birds.

Carson's major achievement in *Silent Spring* was to bring together and synthesize the findings of a wide range of scientists, government agencies, and conservationists, often hidden away in specialist journals, which had been pointing to the environmental dangers of organochlorine pesticides for a number of years. While much of the impact of *Silent Spring* stemmed from its claims about the actual and potential effects of insecticides on human health, it was Carson's account of the consequences for bird populations of the use of these new chemicals that not only bequeathed the book its title, but also strikingly dramatized the dangers they posed to the intricate web of life.

Some of the evidence of bird deaths cited by Carson drew on the observations of her fellow 'Audubonites': the members of the state and national Audubon societies in the United States, who were devoted to the cause of bird conservation and bird study. Carson herself was a member of the (Washington) DC Audubon Society and was part of its network of amateur observers and recreational birdwatchers. Anecdotal evidence from Audubonites and research supported by the National Association of Audubon Societies (NAAS) formed important sources of evidence for Carson's account in *Silent Spring*. Carson was also an acquaintance of Roger Tory Peterson, a leading light within the NAAS as well as one of the most high-profile publicists for American birdwatching and bird conservation. Through Tory Peterson and his close friend James Fisher, the leading British birdwatcher and author, Carson gained access to the emerging evidence from Britain about birds and toxic chemicals. This rooted Carson's writing within the trans-Atlantic networks of bird conservationists and advocates of bird study.

In this chapter I explore the impact of *Silent Spring* in the United States and Britain, placing it within the wider debate among bird conservation organizations, conservation-minded scientists, and policymakers about the scale and pace of environmental change and the consequences of this for wild birds. To this end, the chapter reflects on the way Carson, the NAAS, the BTO, and the RSPB helped to make visible and legible the new threats to birdlife from toxic chemicals and linked these threats to an understanding of a wider crisis of the natural environment within Britain and North America from the early 1950s.

This was a crisis produced by a distinctive concatenation of forces peculiar to the immediate post-war decades. Whereas in the years between 1890 and 1920, when the Audubon societies and the RSPB had first been formed and become active, the main threats to wild birds came, as Carson herself noted, from the 'massacre of the shore birds by the market gunners [and] the near extermination of the egrets for their plumage', post-war development marked 'a new chapter' and a 'new kind of havoc' for birdlife.[9] This included not just the effects of agrichemicals on the environment and the wider intensification of agriculture that were so central to Carson's book, but also the increased development pressure from industry, housing, tourism, and transport and the impact of oil pollution at sea.

The scale of the latter phenomenon was brought home dramatically in the United Kingdom by the sinking of the oil tanker Torrey Canyon off the Scilly Isles in 1967 – one of Britain's worst maritime pollution disasters. The massive environmental impact of the tanker's sinking – caused by the release of thousands of gallons of crude oil and by some of the measures taken to mitigate its effects – not only devastated seabird populations around the British coast, but came to exemplify the way 'economic progress' seemed to have reached catastrophic new levels for other species, especially bird life, and for the fabric of the natural world.

In exploring how a writer like Carson and conservation groups such as the NAAS, the RSPB, and the BTO linked together local, initially unexplained bird deaths and demonstrated how they formed part of a wider national and international phenomenon, the account that follows details the way these social actors built on the principles and practice of bird study that had been undertaken from the early years of the twentieth century in both Britain and the United States. These studies not only provided a crucial benchmark against which to measure the declines in post-war bird populations, but also provided the tools to generate new evidence and new knowledge about the status of bird species. This knowledge circulated across the Atlantic through the long-standing relations between bird scientists and conservationists in the United States and United Kingdom, helping to sustain an expanding field of conservation science, driving the modernization of bird conservation in the United Kingdom, and influencing the reform of public policy towards the environment on both sides of the Atlantic.

In drawing out these international dimensions of the 'toxins crisis' and their effects on wild birds, the chapter also seeks to show how the crisis played out in different ways in Britain and America. Not only were the policies of the US Department of Agriculture and other State bodies different from those pursued by the Ministry for Agriculture, Fisheries and Food and statutory agencies in the United Kingdom, but the avifauna and geography of the United States was also different from that of the United Kingdom.

This meant that the crisis of human–avian relations involved (mostly) different species of birds in different spaces and landscapes. As *Silent Spring* made clear, the US experience of mass bird deaths took place across the distinctive geography of the United States: from backyard, suburban America along the Eastern Seaboard, the Southeast, and the Midwest of the United States to the lakes and forests of 'wild America'. In Britain, it was the fields and hedgerows of Britain's lowland agricultural landscapes that were the principal focus of the crisis, with Britain's coast and inshore waters affected by the separate consequences of development pressure and oil pollution. In this regard, the chapter draws out the different landscapes and spaces in the United States and Britain through which the environmental crisis and its effects on wild birds were played out.

If the avian *dramatis personae* and landscape geography worked to differentiate the British and American experiences of the crisis in wild bird populations, birdwatchers on both sides of the Atlantic were forced to confront similar distressing encounters with dead and dying birds and to wonder where some of their familiar species had gone. The emotional dynamics of the 'toxins crisis' and the other sociogenic environmental threats generated a shared experience and a common bond. In both 'wild America' and its suburban backyards and in the British countryside, the environmental crisis worked to reconfigure human–avian relations. While wild birds continued to evoke great pleasure, joy, and a sense of wonder, this was now mingled with sadness, melancholia, and feelings of loss. Among the RSPB's members in the United Kingdom and Audubonites in the United States, looking at birds became cut across by the shadow of death, by population decline, by distressing encounters with mass bird casualties, and by landscapes transformed by the loss of birds.

This cumulative sense of loss was evident in the letters that NAAS members wrote to their society magazine and in the bodies of birds sent by BTO and RSPB members to their societies, and it was powerfully articulated in the RSPB's campaigning for bird protection in an era of new environmental crisis. In this latter endeavour, the Society drew on the support of the notable nature writer J.A. Baker, whose literary meditation on the threats to birds and Britain's wild places offered powerful testimony about the environmental crisis of the immediate post-war decades and the structuring of human–avian relations. In particular, Baker's award-winning book *The Peregrine*, published in 1967, articulated the dangers posed to birds of prey by organochlorine pesticides and the wider crisis in the British countryside of which this formed a part. While Carson was an important interlocutor for Baker in the writing of *The Peregrine*, his book, like Carson's, stirred his readers to recount their own tales of loss in their relations with wild birds and their sense of the impoverishment of landscapes stripped of the animating presence of charismatic bird species.

AND NO BIRDS SING

In the summer of 1956 the *Audubon Magazine*, the National Audubon Society's members' publication, ran two articles by Paul F. Springer, a US Fish and Wildlife Service biologist. Both were titled 'Insecticides – boon or bane?' Springer noted that his department, along with Audubon societies, received many inquiries each year from 'nature lovers' about whether the use of modern insecticides in agriculture and around the home was, as he summarized it, 'compatible with the maintenance of wildlife'. Drawing on evidence from the US Department of Agriculture – a strong promoter of the benefits of the new chemicals – and that of the Department of Health, Springer put forward the economic and public health case for the advantages of insect control. As he reminded his Audubon Society readers, 'it is well to recall that were it not for insecticides, the production of our food and fibre (cotton) and the protection of the health of ourselves and our animals would be immeasurably more difficult and our standards of living considerably lower'.[10]

He noted, however, that these social and economic benefits came with downsides and dangers. As he pithily put it, 'being poisons, they [the new insecticides] can be harmful to birds, mammals and fish'. These harmful effects went beyond the immediate risk of death for birds and other wildlife that came into contact with the chemicals. Citing research by his department on the effects of insecticides on game birds, he showed that sublethal doses could affect game bird reproduction. The most insidious risk of the new insecticides, he warned, stemmed from their persistence within the bodies of birds, and it was these 'high residual quantities' that posed longer-term dangers to bird health and bird populations. 'Bird watchers,' he went on, 'garden club members and other conservation groups have justifiable concern as to the possible hazards to wildlife from insect control operations.' What was needed, he concluded, was the careful management of the 'problem of insecticides' in order to lessen their effects on the environment.[11]

Springer's articles were an early warning of the dangers of the mass application of organochlorine pesticides in America's towns, gardens, and countryside.[12] It was significant that he worked for the US Fish and Wildlife Service as this agency was the most sceptical of the US government bodies about the benefits of organochlorines, and it was certainly the keenest to research the effects that they might have on wildlife and other 'natural resources'.

Over the next three years, the *Audubon Magazine*, largely prompted by a series of spraying campaigns led by the Department of Agriculture against a range of insects pests, published news report, articles, and letters that documented the catastrophic effects of these campaigns on wild birds.[13] In one response to the emerging problem, the NAAS sought to stop the fire ant eradication programme in the spring of 1958, requesting a pause to the spraying until the Department of Agriculture was able to provide evidence that the chemicals did not do serious damage to wildlife. As John Baker, the president of the NAAS, argued, drawing a link with a concern at the forefront of public consciousness, 'insecticide hazards may well rank in seriousness … with the dangers of radioactive fallout… The cumulative secondary poisoning of human beings and wildlife … may become catastrophic.'[14] This message chimed with *Audubon Magazine* readers like Martin R. Hause, whose letter to the magazine in May/June 1958 echoed Baker's concerns: 'Next to the threat of nuclear extermination,' Hause argued, 'the widespread use of poisons in food and agriculture poses the greatest danger to civilization.'[15]

Articles in the *Audubon Magazine* began to document the dramatic consequences of the mass spraying programmes. Alfred G.

Ether of Michigan captured the sense of assault on his home and its community of birds from the 'thundering spray plane that rocks my house' and the deleterious effects on spraying on his relationship to his familiar birds: 'The drifting spray is already upon us. I close the windows in the house and am glad that the car is in the garage… I listen to the birds in the garden singing: the oriole that is sitting on eggs in the elm … the yellow warbler that I saw yesterday while lying on my back looking up into the leaves… They are kind companions. I am struck with the guilt that I am a part of the destruction wrought upon nature by this poison cloud descending.'[16]

Other dramatic evidence of the new dangers facing birds came from university campuses. At Michigan State University, the zoology professor George Wallace first observed the dead and dying birds found in large numbers each spring on the university campus and in adjacent parts of East Lansing in 1954. He then began to study them. The mass deaths coincided with an intensive spraying programme designed to combat Dutch elm disease and to control mosquitoes. In a paper delivered to the Annual Convention of the National Audubon Society in November 1958 and summarized in the *Audubon Magazine* early in 1959, Wallace reported that over the four years of the programme, American robins had been virtually eradicated from the campus and its environs.[17] With forty-nine other species of summer resident birds disappearing or being sharply reduced in numbers, Wallace concluded that 'it is increasingly evident that robins [and other birds] and an intensive Dutch elm programme cannot co-exist'.[18]

Wallace's evidence was supported by the research of Dr Joe Hickey at the University of Wisconsin. Hickey found that between 68 and 99 per cent of robins disappeared in those communities that were sprayed for Dutch elm disease.[19] The Massachusetts Audubon Society picked up the findings of the two scientists in early 1962, issuing a press release under the headline 'Must we lose our robins?' Alongside photographics of dead robins, the text drew a fateful parallel between the current state of robins and the extirpation of two celebrated bird species: the dodo (from Mauritius) and (closer to home) the passenger pigeon, a once massively abundant species in the United States that had become extinct in 1932. Was the American robin heading the same way, the Massachusetts Audubon Society asked? With evidence from Michigan (Wallace's study) showing a 92 per cent decline in some areas and with sample counts in parts of Massachusetts in 1960 and 1961 indicating an 80 per cent decline on 1939 numbers, the Society feared the worst. It pointed the finger at the 'unwarranted application of new and powerful pesticides', arguing for immediate action to protect 'our wild heritage'.[20]

A few years earlier, records from the Audubon Society magazine *Field Notes* had begun to capture the scale of bird population declines in the American Southeast. In Baton Rouge an Audubonite reported that the contents of her backyard bird feeders had been untouched for weeks. Another revealed that the view from his picture window, once 'splashed with the red of 40–50 cardinals and crowded with other species', now only held one or two birds.[21]

Reports came in from other parts of the country too. The Trailside Museum of Natural History in Illinois revealed that it had received many enquires from residents in the surrounding suburbs who had found sick and dying birds following the spraying of trees for Dutch elm disease. The museum's Virginia Moe noted the typical condition of the birds: 'Symptoms are paralysis and constant trembling and palpitation. When gently held with the legs dangling the toes of the feet move constantly as if the birds were typewriting. Death follows in an hour or so.'[22]

Local newspapers also registered the phenomenon. In July 1958, for example, a letter to Vermont's *Caledonian Record* wondered: 'Are we killing off birds?' The correspondent claimed that since the introduction of DDT to the area to combat Dutch elm disease, 'we have hundreds of poisoned birds delivered to us by local housewives. The toll is indeed alarming.'[23] In February 1963 the *Miami Herald* also reported scores of robins dying at Bay Point. A resident revealed how 'birds have been falling out of our oak tree and we've been picking them up for the past two days'. Another resident revealed that following the spraying of her lawn, 'birds began falling from the trees'.[24]

Alongside the harrowing evidence of the deaths of common garden birds, the *Audubon Magazine* also revealed disturbing evidence about the perplexing decline in one of America's most iconic birds of prey: the bald eagle. Since the late 1930s, retired banker Charles L. Broley, with the encouragement of the NAAS, had studied the Florida population of the birds. Much of Broley's effort was focused on banding or ringing the young eagles in order to learn something about their reproductive success.

By 1946 Broley was banding around 150 young eagles a season. In the late 1940s, however, he began to notice a decline in the productivity of the birds. This decline accelerated through the late 1940s, and by 1950 Broley banded only twenty-four young birds, a decrease of 78 per cent on his 1946 totals. In a 100-mile area of coastal Florida the eagles' breeding success had dropped from fifty-six nests raising 103 young in 1946 to seven nests raising eight young in 1957. Broley concluded that around 80 per cent of Florida's bald eagles were sterile.[25] While he did not have conclusive evidence for why this had occurred, he suggested that Florida's heavy spraying of insecticides must be implicated. Studies of Tampa Bay in the early 1950s had found high levels of DDT residue. As a fish eater, bald eagles were likely to be being affected by eating fish contaminated with the chemicals.[26]

The *New York Times* picked up on Broley's article in *Audubon Magazine* and ran with a dramatic and suggestive headline: 'US is losing its bald eagles; sterility suspected, DDT cited'.[27] The bird had a particularly significant status in American life having been dedicated as the symbol of the American nation by the act of the Second Continental Congress in 1782. As the NAAS president Carl W. Buchheister put it, the eagle was a bird that signified 'our nation's majesty and might'.

Because of its symbolic status and charismatic quality as a big, powerful bird of prey, the NAAS decided to support a new scientific study of the bald eagle across its continental range. Raising $50,000, some of it from the society's members, and working with federal, state, and provincial wildlife agencies, the NAAS began a five-year research programme in 1961.[28] The project also drew on existing records of the successful nesting of bald eagles collected by Audubon society members.

Evidence was additionally taken from another long-running study of birds of prey conducted by Maurice Broun, the curator of the Hawk Mountain Sanctuary in Pennsylvania. The sanctuary had been established in 1934 by wealthy New York conservationist Rosalie Edge. Edge first leased and then purchased 1,400 acres of land along the Kittatinny Ridge to help protect migrating birds of prey from the hawk-gunners that annually took a heavy toll on the birds as they flew through Pennsylvania.[29] Broun took on the role of warden for the sanctuary and began a series of seasonal counts of the migrating birds of prey.[30] His records included the numbers of bald eagles passing past Hawk Mountain, and they proved significant

Figure 4.1 Rachel Carson watching migrating hawks at Hawk Mountain, PA (1945).

in supporting Charles Broley's findings about the reproductive failure of the birds. By the late 1950s Broun noted that young, first-year bald eagles were among the rarest raptors passing through Hawk Mountain.[31]

*

In October 1945 Rachel Carson, then working for the US Fish and Wildlife Service, visited Hawk Mountain with her colleague Shirley Briggs. Briggs photographed Carson perched on North Lookout, an exposed rock promontory, scanning for birds of prey through her binoculars. Carson later recorded details of the landscape and the birds in her field notes. The hawks came, she noted, 'like brown leaves drifting on the wind. Sometimes a lone bird rode the air currents; sometimes several at a time, sweeping upward until they were only specks against the clouds.' Her records also made much of the fact that the sandstone and limestone rocks on which she was sitting had once formed the bed of an ancient ocean. As she noted, 'these Appalachian highlands are reminders of those ancient seas that … once lay over this land'. 'I lie back,' she concluded, 'with half-closed eyes and try to realize that I am at the bottom of another ocean – an ocean of air on which the hawks are sailing.'[32]

Four years earlier Carson had published her first book, *Under the Sea-Wind*, a 'naturalist picture of ocean life' as it was subtitled. Oceans and the ecology of sea life were a central interest in Carson's life and she had trained as a marine biologist. While *Under the Sea-Wind* made little impact on its initial publication, her follow-up book, *The Sea Around Us* (1951), proved to be much more successful. It was a lyrical study of the ocean, marked by Carson's characteristic focus on stimulating a sense of wonder and awe towards the natural world.[33] The book was serialized in the *New Yorker*, chosen as a Book-of-the-Month-Club selection, and became a best seller. Carson followed it up with *The Edge of the Sea* (1955), the third of her trilogy of 'sea books'. By this time the success of *The Sea Around Us* had allowed Carson to resign from her job at the Fish and Wildlife Service and to become a full-time writer. By the late 1950s, she was established as one of America's most well-known and popular science writers – her books familiar fixtures within the homes of millions of middle-class Americans.[34]

It was on the back of this success that Carson began to consider a new, very different kind of book. It would focus on the potential dangers of organochlorine pesticides to American wildlife and human health. Initially titled *Man Against the Earth*, Carson began to collect material in 1957 for her 'poisons book'.[35] Her research files filled with material drawn from a wide array of sources. Significantly, this included material produced by her old department, the Fish and Wildlife Service, as well as news and articles published in *Audubon Magazine* and the society's other publications.

Partly hampered by ill health, Carson's research strategy leant heavily on corresponding with informants who could provide her with some of the information that she needed for her study. One of these correspondents was George Wallace from Michigan State University. In October 1959 Wallace sent Carson a copy of his paper 'Another year of robin losses on a university campus'. In the accompanying notes he wrote: 'Hope this serves as grist to your mill! Don't believe robin losses have been measured in just this way before – more dead than living! We have deliberately refrained from any mention of insecticides... We are trying to get set up for making analyses of the dead birds – proving to be difficult and time consuming.'[36]

Copies of Wallace's articles in the *Audubon Magazine* were also assiduously collected by Carson, and she wrote directly to senior NAAS members asking them for any more information that they might have on insecticides and bird life. One of the correspondents was C. Russell Mason, executive director of the Florida Audubon Association. Carson was particularly interested, as she put it, 'in the problem of the bald eagle'. 'I shall be very grateful,' she went on, 'if you could summarize [the research from the *Florida Naturalist*]. I am always eager for more [material] especially as it relates to local situations.'[37] Russell Mason replied that most of his association's 'eagle work' built on Charles Broley's surveys of nests (Broley had died in 1959). Russell Mason, however, cautioned Carson about the link between the decline of the eagles and organochlorine chemicals. As he put it, 'Broley's claim that sterility in eagles is induced by DDT is not proved of course.'[38]

Carson was pricked by Russell Mason's comment on Broley's studies and wrote back insisting that there was strong supporting evidence from elsewhere to suggest that Broley's inference about the effects of DDT were not misplaced: 'Broley's theory does not seem at all far-fetched when one considers the facts that have been brought out by De Witt and others in relation to quail and by Wallace in relation to robins. With regard to the food chains ... I am sure you have seen the very interesting account by Rudd of the situation in Clear Lake, California, which provides such a significant illustration of the way these chemicals may operate across a number of successive links in the food chain.'[39]

Carson also corresponded with the NAAS's research biologist Harold S. Peters. He was particularly well informed on the spraying programmes of the US Department of Agriculture and local control agencies and offered detailed information on the way the sprays were administered and on their effects.[40] He communicated to Carson the emerging evidence of the catastrophic environmental effect of DDT used to treat Dutch elm disease, citing evidence from Wallace that more than '10 million robins have died. In spite of this the elms aren't being saved.' He ended encouragingly: 'I am sure you are treating some of this in your book.'[41]

Carson was indeed making use of Wallace's material for her book as well as the other information provided by her correspondents in and around the NAAS. As Carson worked on her 'poisons book', her agent suggested that, as with *The Sea Around Us*, material from the new book could be extracted and published in the *New Yorker* magazine. A deal was agreed with the publisher and Carson's three articles for the magazine were published just ahead of the release of *Silent Spring* in the autumn of 1962. The first article, published on 16 June 1962, caused, as Linda Lear notes, 'an immediate sensation'.[42] The three articles fuelled huge interest in Carson's book, and it sold rapidly, spending much of the autumn of 1962 at the top of the *New York Times* best-seller list, with sales passing 106,000 in the week before Christmas 1962.[43] By March 1963, it had sold more than 500,000 copies, been selected for the Book-of-the-Month-Club, and had been published in the United Kingdom and West Germany. French, Swedish, Danish, Dutch, Finnish, and Italian translations also appeared over the course of 1963.[44]

In April 1963 CBS television ran a special one-hour programme in its influential 'CBS Reports' strand titled 'The Silent Spring of Rachel Carson'. Carson's clear and calm exposition of her views in the programme contrasted with both the apparent ignorance of government officials on the issue and the wild-eyed defence of toxic chemicals by Dr Robert White-Stevens, who was there to represent the chemical industry. Most observers agreed that Carson had won the debate and therefore set the terms for the public agenda, and with 10–15 millions viewers watching, *Silent Spring* was given a massive sales boost.[45] Carson's performance and the furore around the book stirred the US government. It published its delayed 'The use of pesticides' report soon after, and while the tone of the report was not as critical of pesticide use as Carson's book, it did give legitimacy to many of Carson's criticisms.[46] From this followed new legislation that banned DDT by the mid-1970s and the other organochlorines by the end of the decade. In 1970 President Nixon established the Environmental Protection Agency to regulate pesticide use and to establish new environmental standards.[47]

A large part of the impact of *Silent Spring* on the wider public stemmed from the fact that it found a readership already sensitized to the dangers of unseen environmental threats, an emerging scepticism about scientific progress, and fears that powerful vested interests in American society sought to hide inconvenient truths from the American people. Cold war anxieties about nuclear fallout in particular fed public anxiety about the dark side of science and the lack of accountability of government and corporate interests.[48]

Silent Spring cleverly played on these fears. It mixed careful scientific fact with the language of mythology and the disturbing tropes of European fairy tales. These hinted at the unforetold consequences of human action, of doom and disaster unwittingly created, and of human arrogance leading to hubris. These themes were powerfully established in the prologue to the book. Titled 'A fable for tomorrow', Carson conjured the story of a fictitious American town struck down by a 'strange blight' – as if, as she put it, an 'evil spell' had fallen upon the town, cattle and sheep became ill and died, children got sick, and the birds disappeared. Where once there had been all life living in harmony, now 'everywhere there was the shadow of death'.[49] Drawing on the evidence she had collected from Audubonites and newspaper and magazine reports, Carson's fable focused on the vanishing birds: 'There was a strange stillness. The birds … where had they gone? Many people spoke of them, puzzled and disturbed. The feeding stations in the backyards were deserted. The few birds seen anywhere were moribund; they trembled violently and could not fly. It was a spring without voices … only silence lay over the fields and woods and marsh.'[50]

This was a blight caused, Carson suggested, not by witchcraft or enemy action, but by the people themselves. Drawing on Greek mythology, Carson suggested that modern science had, through organochloride pesticides, created a 'Medea's robe'. Like the gown created by Medea that brought violent death upon its wearer, so 'systemic pesticides' brought death by the same 'indirection'. They turned plants and animals into 'a sort of Medea's robe' that killed by transforming them into poisons.[51] Elsewhere the book spoke of the 'elixirs of death' that brought not life but destruction to rivers, fields, suburban backyards, and the home itself. Carson turned the militarized language of the pest control programmes and their 'war on nature' against the government agencies to assert that they were laying down a 'barrage of poisons' upon the environment and bringing a 'rain of death' upon the surface of the earth.[52]

As Carson showed, not only did these 'biocides' directly bring death, but they also had further sinister effects. By entering the food chain at sublethal levels they had a deferred effect on those species at the top of the chain, such as Charles Broley's bald eagles or the grebes killed in Clear Lake, California following the control of gnats.[53] In what was perhaps the most powerful aspect of Carson's arguments, she drew on ecological understandings of the interdependence of life to point to the risks to human health from indirect exposure to these 'biocides'. Noting that the new chemicals had already become stored in the bodies of most Americans and that they could be found in mothers' breast milk, Carson forced her readers not only to face the unknown risks to human wellbeing, but to see human life as part of, and dependent on, the wider natural world. Human beings, like Broley's eagles, were part of a food chain, and they were therefore bound to and dependent on the wider natural environment.

While she did not propose a total ban on all the new pesticides, Carson did make a powerful case for greater control of their use and for searching for alternative biological solutions to pest control. She also made a political argument throughout *Silent Spring* that government and commercial interests should be subject to greater democratic accountability and that citizens had the right to experience the joy of nature and wild birds as part of a full human flourishing. As she asked: 'Who has decided – who has the right to decide – for the countless legions of people who were not consulted that the supreme value is a world without insects, even though it be also a sterile world un-graced by the curving wing of a bird in flight.'[54]

A PITIFUL SIGHT

The story that Rachel Carson told in *Silent Spring* was largely an American story, rooted – as David Kikele and Daniel Horowitz have argued – in the landscapes created by post-war American affluence, suburbanization, and technological change.[55] The book drew on evidence from, and evoked the streets and backyards of, small town America and the wider landscapes of 'wild America'. Carson's opening fable had summoned up an archetypal post-war American suburb, close to existing farmland and with its backyards supplied with bird feeders. The selection of these landscapes and places reflected the distribution of the spraying patterns of the US Department of Agriculture and other government agencies and the geography of pest populations in the United States. It also drew from the places in which scientists like George Wallace and Joe Hickey, and committed amateurs like Charles Broley and Maurice Broun and hundreds of Audubonites, had made their observations.

The letters that Carson received following her *New Yorker* articles and after the publication of *Silent Spring* confirmed this mapping of the crisis of both wild and suburban America. They included Florida residents like Philip E. Howard Jr, who wrote to Carson to tell her of the disappearance of local birds: 'My wife and I have noted the scarcity of bluebirds, and I have seen fewer eagles in the last 2 or 3 years than formerly and none at all this winter. At dinner the other evening a lady from Orlando told me of seeing a bird collapse on her lawn as he tried to feed while the sprayers were working. She took it in but it died in a short time.'[56] Sandra L. Showalter from Chicago told a similar story: 'I have noticed that in my city, there are almost no birds in areas which formerly used to flourish with them.'[57] From Urbana, Illinois, Marcus Goldmann revealed: 'The effect of the spraying campaign on the birds was … disastrous. For a number of years robins seemed to have disappeared … and in 1963 their numbers were … much smaller than in the years before the spraying.' In Minnesota, another part of the Midwest, A.O. Hage told Carson: 'Years ago here in the Red River Valley … there were many kinds of birds… Now they have almost vanished.'[58]

Among the letters sent to Carson via Houghton Mifflin, her publisher, offering tales about the disappearance of familiar birds were some from Europe and further afield. One, from Harry R. Lillie, an acquaintance of Carson's in Scotland, spoke of the situation in the United Kingdom. He quoted James Fisher, 'our ornithologist', as he chummily put it, who had praised Carson's book in his BBC radio series 'Birds in Britain'. Lillie explained to Carson how Fisher, picking up on the BTO's survey of peregrines, had drawn attention to the 'losses of our hawks from eating insecticide poisoned birds'. 'Our peregrine population,' he revealed, 'had been greatly diminished, down from 750 pairs before the war to 135 sites today.'[59]

Carson did not need reminding of the British evidence, however. *Silent Spring* drew on research into the effects of toxic chemicals in the United Kingdom, including early evidence from the BTO peregrine study. These sections of her book brought a picture of the British countryside into the largely American geography of loss that she documented. Carson knew of James Fisher not only through his writings and appearances at NAAS conventions, but also through meeting him through their mutual friend Roger Tory Peterson.[60] Fisher was also one of the sympathetic potential reviewers that the NAAS suggested Houghton Mifflin should send pre-publication copies of *Silent Spring* to. These connections, part of the transatlantic network of contemporary ornithology, enabled *Silent Spring* to enjoy a rapid circulation within the United Kingdom, and they meant that the book also fed into an existing debate within conservation and government circles about the potential dangers of new agriculture insecticides. In the next section I explore the British response to both the emergent 'toxins crisis' and to *Silent Spring*, focusing on the moves by conservation organizations to document the crisis and find scientific explanations for its causes.

'AN ECOLOGICAL TRAGEDY'

The RSPB and the BTO had both independently begun to investigate the issue of toxic chemicals and their effects on birdlife from the mid-1950s. In 1955 Lord Hurcombe, chairman of the RSPB Council, had identified in the Society's members' magazine the 'often unscientific use of insecticides and herbicides' as one of the 'adverse forces' threatening birds and the wider 'natural heritage' of the United Kingdom.[61]

Four years later, the BTO established a Toxic Chemicals Group under the direction of its Scientific Advisory Group. This group reviewed the American and (smaller) European literature on the science and uses of toxic chemicals, actively communicating with both the NAAS and the US Fish and Wildlife Service.[62] In 1959 the RSPB, largely through the endeavours of its assistant secretary Peter Conder, had begun to explore news reports of animal and bird deaths.[63] At the suggestion of the RSPB, the enquiries of both societies were linked together and a joint BTO–RSPB committee on toxic chemicals was established in July 1960. The BTO's Stanley Cramp chaired the committee, with Peter Conder acting as secretary.[64]

The joint committee's first report, issued in April 1961, covered the deaths of birds and mammals for the first half of 1960. Most of the published evidence related to deaths caused by toxic seed dressings. Among the sixty-seven reported incidents was evidence from a Yorkshire landowner and his keeper, who, across two weekends, found 182 birds of eleven different species. Many of the birds, like those also recovered from a garden in Kent adjacent to agricultural land, were described as 'standing about shivering'. Others were described as 'flapping around and gasping', 'floundering', and 'staggering about'.[65]

The report conceded that the joint committee's initial analyses of the causes of bird deaths were inconclusive, given the problem of detecting small residues of poison in moribund birds and the complexity of the chemical compounds involved. Nonetheless, the report noted that most of the birds that were found dead or dying had been killed not directly by the poisons but probably from eating contaminated insects. Looking for corroboration from across the Atlantic, Cramp and Conder suggested that the 'American experience' documented in the work of Wallace and Hickey had shown how these chlorinated hydrocarbons killed in an insidious way through the accumulation of sublethal doses in the bodies of birds.[66]

The second report of the joint committee continued to draw on this US evidence to point the finger of blame at toxic chemicals for the incidents of mass bird fatalities in the United Kingdom. The report mapped the incidents of birds' deaths nationally, showing in diagrammatic form their concentration in those parts of Britain dominated by arable farming. It also captured a big increase in 'kills' from the first six months of 1961. These had reached 347 incidents – a fourfold increase on 1960 records – with the vast majority still associated with toxic seed dressings. The most dramatic incident came from Tumby in Lincolnshire, where – on a large, mostly arable, estate – the corpses of 382 birds were collected between March and June. Survey counts from Tumby suggested that the total mortality for the four most affected bird species on the estate was likely to have been huge. They suggested that for the wood pigeon, the commonest victim, more than 5,500 birds were likely to have died.[67]

The dramatic scale of the Tumby incident, coupled with other mass 'kills' across eastern and southern Britain, caught the attention of the broadsheet press and gave a new level of visibility to the problem of toxic chemicals for wildlife.[68] In this regard, 1961 represented an *annus horribilis* for Britain's seed-eating country birds. The dramatic scale of the mass bird deaths caught the eye of Rachel Carson and, drawing on the BTO–RSPB report, she cited the Tumby event in *Silent Spring*. In the third, fourth, and fifth reports of the joint committee, the number of bird deaths remained very high, with new evidence suggesting that toxic sprays and horticultural pesticides were increasing as causes of death relative to those from seed dressings. Despite the voluntary ban on seed dressings for spring-sown crops, however, more than half the bodies of birds analysed in the mid-1960s continued to involve aldrin, dieldrin, and heptachlor, the main (and highly toxic) chemicals used in seed dressings.[69]

The BTO–RSPB joint committee was increasingly exercised by the growing problem of population declines among birds of prey, particularly in those counties where toxic seed dressings were being most heavily used. The fate of the peregrine was especially troubling for the committee, with evidence from Cornwall indicating a 'catastrophic decline' and the finger of suspicion again pointing towards the bird being 'another indirect casualty of toxic chemicals'.[70] The interim report of the BTO's peregrine study, released in June 1962, confirmed earlier fears about the crash in its population, particularly in Southern England, while the evidence cited by Carson in *Silent Spring* and findings from Roger Tory Peterson about the poor breeding success of ospreys in the Connecticut River in the eastern United States were used by the BTO–RSPB committee to underline the links between toxic agrichemicals and the population decline of birds of prey on both sides of the Atlantic.[71] As the committee's fifth report gravely suggested, citing further North American research, the possibility of losing 'one or more of our predatory species … is not an idle fear as shown by the results of an intensive census of peregrines in eastern Canada and the United States in 1964 which failed to disclose a single occupied breeding cliff'.[72]

The focus on birds of prey was telling. Mirroring the valuation of raptors in the United States, especially the bald eagle, as symbols of wild America and embodiments of the national character, birds like the peregrine (despite its cosmopolitan distribution) were seen as part of Britain's 'national heritage'.[73] For conservationists, their decline subtracted something essential from the landscapes that they had inhabited and helped to animate. While other common bird species could also invoke this sense of place, it was the pull of charismatic birds of prey that most riveted many conservationists and birdwatchers.

Through the worst period of mass bird deaths in the early 1960s, most birdwatchers in Britain did not routinely encounter dead and dying country birds. This meant that the suburban American experience described by Carson was not repeated in the United Kingdom. It was, however, the decline of birds of prey and the absence for birdwatchers of the typically rare and fleeting encounters with these birds in wild landscapes that carried a more tangible sense of loss.

For the BTO–RSPB joint committee, the records of the declines of birds of prey that their research had begun to reveal was not matched by reliable wider data on the national populations of these birds, save for the peregrine. As the BTO's Stanley Cramp noted, there were 'too few accurate census figures [of most of our birds of prey] either before or since the new synthetic pesticides begin to be used on a large scale'.[74]

Some attempt at an estimate for one of the two most common raptors, the kestrel, was made by Richard Fitter, for the Council for Nature, in 1963 by gathering information from observers in England and Wales. Cramp himself used county bird reports, and their records from amateur observers, for 1960 and 1961 to confirm the general picture for sparrowhawks. The BTO's nest record scheme also provided data on the populations of both species.[75] Within the BTO, however, there was a recognition that the Trust needed to develop a new systematic approach to bird populations. In 1962 its council agreed to refocus the priorities of the Trust, placing emphasis on the study of bird populations and the ecological factors affecting them.[76]

An important impetus for this policy initiative came from Nature Conservancy and its request for a new enquiry, to be led by the BTO, into common farmland birds in the wake of the emerging evidence about toxic chemicals. Given the title the Common Bird Census, it aimed to establish basic data on population fluctuations among Britain's most common birds. Methodologically, the census counted breeding birds in a number of sample areas, with the counts repeated annually.[77] Significantly, the initial focus was on agricultural land. As the census's lead investigators Kenneth Williamson and R.C. Holmes suggested: 'It is on farmland that bird life is perhaps most seriously threatened; the replacement of hedgerows by wire fences and the growing use of toxic sprays and seed dressings, have raised the grave suspicion of a widespread decline in bird populations.'[78]

The first Common Bird Census counts were made during the breeding season of 1962. Their focus on measuring the link between farmland bird populations and agrichemicals was, however, disrupted by the severe winter of 1962/63. As Williamson and Holmes noted, this was the most prolonged cold spell in Britain since 1740, with the average temperature in lowland England being zero degrees centigrade between December 1962 and February 1963.[79]

The extreme weather caused mass bird fatalities. A study published in *British Birds* recorded that nearly 4,000 wood pigeons had been found dead between November 1962 and March 1963, with nearly 1,500 starlings also found dead over the same period.[80] The effects of the winter weather on birds was so pronounced that the spring 1963 breeding census was unable to 'isolate any other influences' on bird population declines.

Breeding bird numbers for the more common farmland birds recovered to their 1962 levels by 1966, and the BTO took the decision to treat 1966 as the 'base year' against which fluctuations in population levels would be compared.[81] By this time, however, the census of kestrel, sparrowhawk, and peregrine populations (the latter of which had been subject to an annual sample census) showed stabilization in numbers. By 1964, certainly, Derek Ratcliffe suggested that the decline of the peregrine had been halted.[82]

If these census results suggested a certain resilience of farmland birds and birds of prey to changes in the environment, as well as the amelioration of the worst effects of toxic chemicals through the government's incremental regulation of them, other environmental threats to wild birds punctured any sense of optimism among bird conservationists. The sinking of the Torrey Canyon oil tanker in March 1967 off the southwest coast of Britain made clear the continuing danger posed to bird life by modern commercial practices. More than 7,500 birds were taken from the sea as a result of the spillage, with more than 2,000 found dead or dying and many of those recovered alive later dying from oil ingestion.[83] Television audiences and newspaper readers were presented with pitiful and disturbing images of dead or dying seabirds covered in crude oil washing up along the coast.[84] Shortly afterwards, in 1969, a further threat to birdlife came with the plan for a third London airport at

Maplin Sands and Foulness Island in Essex, and with it the potential destruction of important wildlife-rich habitats.[85]

These new threats to wild birds, coming so soon after the toxins crisis, prompted a sustained response not only from bird conservationists but also from contemporary nature writing.[86] J.A. Baker's work was especially notable in this regard. His award-winning first book *The Peregrine,* together with an essay that he wrote for the RSPB magazine (the latter of which was also the basis for an RSPB film), articulated a powerful sense of nature in crisis and of the emotional costs of this assault on the countryside and its bird populations.[87] Baker's writing formed a counterpoint to Carson's *Silent Spring,* representing a defiantly local, British literary rendering of her apocalyptic narrative of American loss.

WILDERNESS IN NOT A PLACE

The Peregrine was J.A. Baker's first book. Published in 1967, it won the prestigious Duff Cooper Memorial Prize and was well received in the press and by broadcasters. For John Moore, who had reviewed the original manuscript for the book's publisher, Collins, the book was 'something quite exceptional in the way of nature writing'. The editors of *Audubon Magazine* were also sufficiently taken with the book on its first release to publish extracts from it in the magazine in the autumn of 1967.[88]

Baker's book documented an unnamed narrator's pursuit of wintering peregrine falcons in the English lowlands. Presented in the form of a diary of one year's watching, the book offered intense observations on the life of these wintering falcons and the other bird species found in this corner of the British countryside. It drew heavily on Baker's own field notes, accumulated across ten winters of watching. Although *The Peregrine* contained no place names to identify the landscapes it described, the book was based on Baker's exploration, usually by bike, of the fields, lanes, and woodland east and west of Chelmsford in Essex, along the river valley between Chelmsford and the Essex port of Maldon, and along the north coast of the River Blackwater and the Dengie Peninsula in Essex.[89]

Baker more explicitly eulogized the Essex landscape in his essay 'On the Essex coast', which was published in the RSPB's members' magazine in 1971.[90] The essay featured in an issue of the magazine dedicated to the Society's campaign against the proposed new

Figure 4.2 J.A. Baker on the Essex coast (1950s).

London airport in the Thames Estuary. Baker's essay poetically yet firmly set out the value of this undervalued and threatened environment. Describing the Essex coast, he suggested: 'There is something here, something more than thousands of birds and insects, than the millions of marine creatures. The wilderness is here. To me the wilderness is not a place. It is the indefinable essence or spirit that lives in a place… It is rare now. Man is killing the wilderness, hunting it down. On the east coast of England this is perhaps its last home.'[91]

The palpable threat to this landscape was brought home by Baker's description of a dead seabird – a red-throated diver – washed up on the shoreline, killed by oil pollution. Invoking the recent events of Torrey Canyon and the wider problem of routine oil discharges at sea polluting the British coast, Baker saw the bird as a victim of 'our modern barbarity', dying 'slowly and horribly in a Belsen of floating oil'. If this 'holocaust' of pollution already threatened the Essex coast, then, for Baker, the proposed airport would finish it off: 'Cordoned by motorways, overshadowed by the huge airport city, the uniqueness of the place will be destroyed as though it had been blown to pieces by bombs.'[92]

Readers of the essay in *Birds* were stirred to write to the RSPB. Mrs Molly Drake, from Essex, was particularly taken with Baker's 'eloquent and haunting elegy to the wilderness about to be destroyed'. His essay chimed with her own sense that 'to birdwatch at Foulness in these days is heart-breaking, with the shadow of destruction hanging over it, but we are going ahead with plans for this season's wildfowl counts. Please help us to save what can never be replaced.'[93]

Much of the power of Baker's essay stemmed from the way it conveyed a love for a landscape borne of an intimate knowledge. It revealed his deep attachment to the countryside and the coastline of southeast and mid-Essex, which formed a central element of *The Peregrine.* That book, however, was also a lament to a declining bird and a changing landscape. Like his essay, it was shot through with images of death and dying wildlife, and of the destructive power of human society. Part of the reason for this is that the book reworked Baker's diary entries from the harsh winter of 1962/63. Many of his disturbing encounters with dead and dying birds relate to those killed by the freezing conditions. But the book also centrally documents the effects of toxic pesticides on his birds of prey. As he powerfully records relatively early in the book, in an elegiac reflection on the likely disappearance of the falcons that he was observing so closely, 'few peregrines are left ... many bird on their backs ... withered and burnt away by the filthy insidious pollen of farm chemicals'.[94] Here was a direct reflection on the toxins crisis, on the BTO's peregrine study, and on the moral outrage of *Silent Spring.* For Baker, 'man's' complicity in the destruction of nature, and especially of bird life, was also despairingly invoked much later in the book. Encountering two different species of dead or dying birds, the narrator laments: 'We are killers. We stink of death. We carry it with us. It sticks to us like frost. We cannot tear it away.'[95]

Many of the readers of Baker's *The Peregrine* who wrote to him confessed to the intense emotional impact that his book had had on them. In particular they were moved by the plight of the peregrines he described and by the wider crisis of the countryside that he recorded. Many had never seen peregrines and feared they never would because of the long residual life of toxic chemicals. Many spoke of a feeling of loss, of 'a new heartbreak' at the decline of peregrines, of 'man's onslaught upon the countryside and wildlife', of longing to see the peregrine.[96]

Letters written from the United States and Western Europe shared their common experiences of the decline of birds of prey and the widespread effects of changing agricultural practices.[97] As the American C.H. Shafer put it to Baker: 'We have the same problem that you have with pesticides and other short-sighted idiocies. It remains to be seen whether we will have paved the entire countryside before we get round to putting an end to ourselves.'[98] Other British correspondents reiterated the concern. John Cartright lamented: 'I have just read your book. It stirred me deeply… I have never seen a peregrine and considering the long residual life of DDT the likelihood that I will shrinks with each passing season. A weight lies in me over my share in the guilt all men must share.'[99]

For his readers, then, as for the narrator of *The Peregrine*, the countryside around them had taken on the character of 'a dying world', marked by the loss of charismatic and compelling birds such as the peregrine.

CONCLUSION

Are we being sentimental when we care whether the robin returns to our dooryard or the veery sings in the twilight woods? … A world that is no longer fit for wild plants, that is no longer graced by the flight of birds, a world whose forests are empty and lifeless is not likely to be a fit habitat for man himself, for those things are the symptoms of an ailing world.

Rachel Carson, *Audubon Magazine* (September/October 1963, 263)[100]

Rachel Carson's response to her critics was published in the *Audubon Magazine,* and like J.A. Baker's requiem for the peregrine and the wild places of England, it sought to stress a new ethic of human flourishing – one rooted in an ecological understanding of the natural environment and an assertion of the value of 'more-than-human worlds'. Her sustained and careful attack on the environmental costs of the new post-war 'biocides', including their consequences for wild birds, had mobilized amateur observations as well as scientific studies to dramatize the assault on nature and the web of life itself from the toxic agrichemicals.

This chapter has sought to place Carson's interventions within the wider network of bird conservation organizations and scientific ornithology on both sides of the Atlantic, and to show how this coalition of amateur birdwatchers, the investigations of their societies, and the support of sympathetic scientists within government agencies and universities were able to give political visibility to the toxins crisis. Records of dead and dying birds were not only a central feature of the 'rain of death' brought by organochlorines on American suburbs, parts of 'wild America', and the British countryside, but were also transformed by Carson and others into both useful hard data and dystopian imagery in the campaign against the 'biocides'.

The campaigns had some success in pushing policymakers on both sides of the Atlantic to limit the use of, or ban, the new toxic agricultural chemicals. These moves happened more quickly in the United States, with DDT effectively banned by the mid-1970s. In the United Kingdom the use of these toxic chemicals was progressively reduced through the 1960s, with the European Union finally banning them in 1986.[101]

These important gains, however, were not able arrest the 'engine of destruction' upon the natural world. With the further intensification of agriculture in Britain, the numbers of farmland birds declined by between 50 per cent and 90 per cent through the 1970s and 1980s.[102] Birds of prey, however, did undergo a slow and sustained recovery once persistent organochlorine pesticides had been removed from the environment. The adaptability of these birds – particularly the osprey and bald eagle in the United States and peregrines in the United Kingdom – constituted success stories for conservationists and revealed the capacity of these birds to adapt to environmental change and to exploit new opportunities within human-made environments.

One consequence of the success of *Silent Spring* and the continuing pressures on common birds from sociogenic environmental change was that conservation organizations like the NAAS and the RSPB witnessed dramatic increases in their memberships. In 1966 the NAAS saw 8,000 members join in one year, taking its total membership above 45,000 members – up from 30,000 in 1959. The RSPB's membership also rose sharply in the 1960s, growing at an annual rate of nearly 4,000 a year in the late 1960s, and booming by nearly 20,000 in 1970 alone to stand at over 65,500 members.[103]

Their success in making visible and legible to the public the new threats to birdlife contributed to the growing size and financial muscle of the bird protection societies. Despite its more modest membership growth, the BTO, as we have seen, placed national population studies at the heart of its refocused approach to bird study. In the early 1970s, the Trust's new director, Jim Flegg, set out a vision for the BTO that emphasized the role of birds as 'the most sensitive biological indicators of environmental change'. Understanding the link between bird populations and the changing environment became the principal focus of the Trust under Flegg's leadership. At the heart of this was a focus on the defence of 'the ecological diversity' that he saw as central to thriving bird populations and a healthy environment.

One of the key BTO projects that carried these ambitions was the creation of the first national breeding bird atlas for Britain. This drew on the records of more than 10,000 contributors and represented the first national mapping of Britain's breeding birds. Based closely on the the Botanical Society's *Atlas of the British Flora*, the *Atlas of Breeding Birds* used the same ten-kilometre square national grid to capture bird populations.[104]

The RSPB – buoyed by its big surge in members, and led from 1965 by Peter Conder, that 'bonny fighter' in the toxins battles, as a colleague described him – embarked on a professionalization of its head office. This included appointing a new staff biologist, a reserves manager, an education officer, a films officer, a sales officer, and a development officer to continue to grow its membership.

A new reserves acquisition programme was central to this modernization of the organization and its approach to bird conservation. Writing in *Birds* to launch the new reserves appeal in the summer of 1967, Conder drew a strong connection between the toxins crisis, the Torrey Canyon disaster, and the need for new reserves.[105] This focus on acquiring new land for bird reserves became central to the RSPB's approach to bird conservation through the 1970s and to how it sought to shape the relationship between wild birds and its members and others visitors. Reserves like the society's flagship site on the Minsmere Levels in Suffolk, as we saw in the last chapter, offered close encounters with wild birds that were increasingly difficult to engender in the wider countryside. The relative abundance and concentration of birds in managed landscapes like that of Minsmere brought home the loss of birds in the wider countryside for many visitors. Even in these protected spaces, however, the shadow of loss coloured the experience of watching birds. To look was to see if not a dying world, as Baker suggested, then a set of landscapes in which the formerly abundant common birds were in steep decline. It would bequeath a new set of experiential norms for watching birds in the era of the new environmental crisis.

5

Wildfowling and Sporting Naturalism

The true philosopher of the gun is the wildfowler, for he must have the sensitive eye of an artist, a love of solitude and lonely places. He measures beauty by the flash of a bird's wing, by the glint of dawn on sliding waters, by the march of slow clouds. He is the son of solitude, the lonely one.

James Wentworth Day, *Wild Wings and Some Footsteps* (1948, 193)

I still look back and wonder just how different my life would have been had I never been a fanatical wildfowler.

Peter Scott, *The Eye of the Wind* (1961, 104)

It is in October that one may see the skeins of geese arrowing their way towards the south… I can never see it unmoved. Primeval forces are there, made for a moment visible.

Nan Shepherd, 'Wild Geese in Glen Callater' (1959)

On 7 January 1962 a meeting was held at the Rose and Crown Hotel in Wisbech, Cambridgeshire to consider the designation of up to two wildfowl refuges or reserves on the Nene Washes and the Ouse Washes. The meeting, chaired by Max Nicholson, the director-general of Nature Conservancy (NC), brought together representatives of local wildfowling associations with members of the Conservancy's Wildfowl Conservation Committee (WCC).

Opening the meeting, Nicholson drew attention to the intense 'pressure of development' upon the Nene and Ouse Washes. This was a long-standing problem linked to the drainage of the fens, but it had accelerated during and after the war through the extensive forms of agricultural improvement of the land and the ploughing up of areas of scientific interest. Referring to the close relationship between sport and conservation and the need to support the international protection of wildfowl, Nicholson urged all those present to support the designation of 'one or more refuges now'. Dr Storey, a senior member of the Wildfowlers' Association of Great Britain and Ireland (WAGBI), reiterated this message, reassuring the wildfowlers present that the creation of these refuges was not 'solely to prevent shooting [but] was intended to benefit wildfowlers by holding birds in the area where they might not otherwise remain'. Those present unanimously agreed to endorse the designation of two refuges.[1]

The meeting in Wisbech built on the initiatives of Max Nicholson and others to encourage dialogue and cooperation between wildfowlers and bird conservationists in the wake of divisions having emerged between these groups during the passage of the Protection of Birds Act (1954). Since 1955 Nicholson had organized informal 'tea-parties' at the NC headquarters in Belgravia, London that aimed to encourage the building of bridges and the fostering of mutual understanding between the two sides. These had developed into the NC's Wildfowl Conservation Committee, which had formed in 1960.[2]

The 'spirit of tolerance and compromise' that the NC and the WCC sought to foster built on the shared desire of wildfowlers and conservationists to protect the numbers of wild geese and ducks breeding and wintering along Britain's coasts and in wetlands, and to see Britain's proposed wildfowl refuges as links in an international chain of protection established across Europe.[3] Central to NC's vision of wildfowl refuges, however, and underpinning the convergence of interests, was the idea that protected spaces for wild birds would coexist with controlled and regulated shooting. 'Rational conservation' and 'legitimate sport' were to be reconciled, with organized wildfowling cast as the friend and not the enemy of bird protection. As the Earl of Mansfield, the president of WAGBI, put it, the 'lunatic fringe' of both pastimes – what he termed the 'selfish gunner' and the 'dear little dickie-bird sentimentalist' – were to be marginalized in the forward-looking alliance of sport and bird protection championed by the representatives of organized wildfowling and conservation.[4]

In this chapter I explore the relationship between wildfowling and conservation, tracing the debates between wildfowlers, bird conservationists, and legislators from the 1920s through the conflicts around the Protection of Birds Act of the mid-1950s and into the 1970s. This was a period during which wildfowling – the shooting of wild ducks, geese, and some species of wading birds – was changed by the growing dominance of the pursuit by sporting amateurs. On the back of rising living standards and increased mobility and leisure time, particularly in the post-war years, these amateurs came to represent the modern face of wildfowling. In doing so, they eclipsed the older traditions of both the professional fisherman-fowlers, who had eked out a living shooting ducks and geese for the market, and the gentlemen shooters of the nineteenth century.[5]

Pressure on coastal estuaries and wetlands from drainage, agricultural intensification, industrial development, and recreational use also threatened the sporting landscapes that these wildfowlers had helped to shape since the eighteenth century. Coupled with attacks from anti-field sports campaigners and the rising tide of bird protection, wildfowlers mobilized to defend what many saw as the traditional rights to free shooting.[6] One dimension of this defence of their sport was the formation of the first national body representing the interests of wildfowlers: the Wildfowlers' Association of Great Britain and Ireland.

Formed in 1908, WAGBI had three principal aims: to support the dwindling number of professional fowlers; to protect wildfowl habitat from drainage and development; and to defend the sport from those who sought to ban or curtail it.[7] In this latter endeavour, WAGBI's chief target was what *Country Life* called the 'growing body of hyper-humanitarian opinion'.[8] WAGBI remained a tiny voluntary organization until well into the post-war period but was given energy and purpose from the mid-1920s by its campaigns to resist what it saw as anti-wildfowling elements within the various wild birds protection bills that came before Parliament.[9]

WAGBI's cause was greatly assisted by a vigorous sporting press and by a number of notable propagandists for the sport, including James Wentworth Day, the most prolific and literate advocate of wildfowling. In the pages of the conservative countryside magazines *The Field*, *Shooting Times*, and *Country Life*, as well as in his many books, Wentworth Day celebrated the history of English wildfowling, its forms of sporting adventure, and the seductive appeal of wild ducks and geese and the environments they inhabited.

In the first part of the chapter, I explore Wentworth Day's writings, reflecting upon how he elaborated the relationship between wildfowlers and wild birds. In doing so, I bring Wentworth Day's arguments into a dialogue with other literary explorations of wildfowling from the inter-war and early post-war years. These include the youthful writings of Peter Scott, a sportsmen, wildfowler, artist, and leading conservationist of the post-war period. Scott's writing, together – crucially – with his wildfowl painting, opened up the pleasures and adventure of wildfowling in a way that had close parallels with Wentworth Day's accounts, and I draw out their common, as well as their sometimes-divergent, insights into the culture of wildfowling in the 1930s, 1940s, and early 1950s. Wentworth Day and Scott remained – and, in the case of Scott, grew as – significant voices in the debates about wildfowling in the immediate post-war years, and both are central to the dialogue between conservation and wildfowling in the mid–late twentieth century.

In the second part of the chapter I explore the exchanges between wildfowlers and bird conservationists that erupted around the Protection of Birds Act (1954). Scott, through the formation of the Severn Wildfowl Trust in 1946 (later renamed the Wildfowl Trust), was the subject of much hostility and criticism from wildfowlers during this period. He played a key role, however, with his close associate Max Nicholson and leading figures within the RSPB, in fostering a dialogue with wildfowling interests. This dialogue was facilitated by a key clause in the RSPB's Royal Charter that precluded the organization from expressing a public opinion about field sports.[10] It also owed much to a relaxed attitude towards the controlled shooting of ducks and geese from those at the top of British bird conservation. It was further underpinned by the elite backgrounds of those at the top of both bird protection and shooting, and by a crossover in membership of those at the upper echelons of the RSPB, the Wildfowl Trust, and WAGBI. These latter connections spread into the sporting press. The RSPB's former secretary and head of its Watcher's Committee, Philip Brown, became first the deputy editor and then the editor of *Shooting Times* in the early 1960s, helping to bring wildfowling and bird conservation closer together on the pages of the paper.

In charting the relationship between wildfowling and bird conservation, the chapter seeks to understand the emotional, aesthetic, and visceral passions for wild ducks and geese forged within the culture of wildfowling and to apprehend the shaping of the fowler within this form of sporting recreation. The latter was variously represented as the 'modern wildfowler' in the first half of the twentieth century, and later restyled under the auspices of WAGBI as the 'new wildfowler' of the 1960s and 1970s.

For all its common cause with conservation across this period, wildfowling involved different kinds of attention to wild birds from the new birdwatchers. It was shaped through distinct, if related, technologies and codes of honour, and, as David Matless has put it, by contrasting 'acts of love and death'.[11] Understanding this distinctiveness is central to the aims of the chapter. At the same time, however, in the account that follows I draw out the ways in which 'legitimate fowlers' and conservation-minded birdwatchers inhabited similar 'animal landscapes' and shared enthusiasms or passions for birds ordered through related cultural practices. These included the active pursuit of vivid and dramatic encounters with wild ducks and geese, often in compelling and beautiful landscapes in demanding weather conditions at 'antisocial' times of the day. Their shared pursuit of wildfowl required some of the same techniques of concealment and stealth, the same masking of the human form in the landscape, and even the adoption of similar types of dress. They shared technologies like field glasses and the use of hides, with wildfowling bequeathing managed environments like the decoy pond and its reed screens to the bird ringer and the bird conservationist. These commonalities of cultural practice, outlook, and sentiment existed even as the two pursuits developed competing moral claims about wild birds and as the sporting and conservation landscapes that they each helped to shape became, by the late 1960s, increasingly incommensurable spaces for encountering wild birds.

THE MODERN FOWLER

As WAGBI members gathered in London in the summer of 1925 to plan their response to the proposed Wild Birds Protection Bill, James Wentworth Day (1899–1983), a journalist working for Lord Beaverbrook's *Express* group of newspapers, stepped forward to become part of the committee set up to secure amendments to the bill. By the autumn he had become the honorary London organizer for WAGBI.[12]

Wentworth Day would become a leading voice in support of wildfowling, using his journalism and editorial positions on papers like *The Field*, *Country Life*, and *Shooting Times* to promote and redefine the identity of the 'modern fowler'. He was a prolific author, but also a combative and controversial figure even within the conservative world of Fleet Street and the sporting press. A supporter of High Tory causes, and with extreme right wing views, he stood unsuccessfully as a Conservative parliamentary candidate (in 1950 and 1951) and acted as the personal representative of the pro-fascist Lady Houston in the mid-1930s, buying and editing the *Saturday Review* for her.

Like Lady Houston, Wentworth Day was impressed by Mussolini but critical of German fascism. In the early 1950s he defended Imperial forms of Britishness and was hostile to migration from the Caribbean. He was author of a host of books on royalty, sport, speed, natural history, travel, and ghosts, including biographies of King George V, the Queen Mother, and Sir Malcolm Campbell. Like many right wing figures of the early–mid twentieth century, including the *Daily Mail* proprietor Alfred Harmsworth, Wentworth Day was fascinated by speed, combining this enthusiasm for modernity with an antediluvian passion for the English countryside. Wentworth Day described his hobbies in his book *Sporting Adventure*, published in 1935, as 'shooting, hunting, sailing, natural history [and] avoiding left wing intelligentsia'.[13] It was this latter book, the writing of which was prompted by a suggestion from Lord Beaverbrook, that formed one of Wentworth Day's earliest journalistic–literary reflections on the world of free shooting and wildfowling – reflections that would grow into a series of books and essays published between the late 1930s and the mid-1970s.

Wentworth Day's writing mixed history, reportage, and personal testimony to uncover and celebrate the traditions of (largely) English wildfowling and the sporting landscapes of private estates, as well as the free shooting of the intertidal zones of the Crown foreshore in which fowling additionally took place. His accounts were marked by a deep knowledge of locality and place and by his ability to present a vivid and intense impression of being in the watery worlds of wildfowl along the English east coast.

Wildfowling occupied only part of the narrative of *Sporting Adventure*, with Wentworth Day also giving a month-by-month account of the pleasures and dangers of 'the countryman's occasional escapes from London', recording days and weekends snatched from the world of journalism.[14] In *The Modern Fowler*, first published a year before *Sporting Adventure*, however, Wentworth Day offered an extended guide to the principles and purpose of 'modern fowling' and where to pursue the sport. The book elaborated on the guiding values of fowling, aligning these with a commitment to natural history.

One dimension of this latter concern was evident in Wentworth Day's criticism of duck decoys, particularly those sponsored by the Dutch state.[15] While duck decoys were in terminal decline in the United Kingdom, those in Holland were flourishing. They provided dead wildfowl for London stores and the wider UK market, and Wentworth Day contended that around 20 million birds a year were killed in Holland. This loss of birds – many taken during the breeding season and involving wildfowl that would otherwise have wintered in Britain – threatened the stock of ducks that formed the basis of the British wildfowlers' sport. Reiterating an argument that he had made at a WAGBI meeting in London in 1925, Wentworth Day pressed for the banning of these imports and for greater protection for wildfowl in northern Europe.[16] In making his case for protection, Wentworth Day warmly praised the American system of wildfowl preservation and the enforcement of strict bag limits throughout the United States.

Ensuring the continuation of what he called 'the finest sport obtainable with the gun', Wentworth Day also drew a number of other targets into his field of vision. Inter-war development, in the form of industry, housing, and road building, was one of these as it threatened formerly rich wildfowling areas. Haunts in southeast Essex that Wentworth Day had known before World War I had been transformed for the worse by this development. What had once been, in the case of Canvey, 'a splendid place for fowl' was now, he opined, 'a trippers' bungalow colony, a nest of huts, shacks, small houses, cheap dance halls and young men in flannels'.[17] Other vulgar interlopers in the Maldon area supplemented these unwanted modern intrusions. Along the River Crouch, a formerly good wildfowling area had been ruined for Wentworth Day by the disturbance of 'cockney gull-shooters … who pursue anything in feathers with motor boats and artillery of all bores'.[18]

The drift of modern life had also negatively affected Wentworth Day's favoured haunts in North Norfolk. Blakeney, on the coast, was now 'overrun with golfers, artists and women in Harris Tweeds', and the National Trust nature reserve at Blakeney Point had been captured by 'the camera-carrying maiden lady type of bird protectionist'.[19] Differentiating his own natural history from these modern 'bird protectionists' and 'bird disturbers', Wentworth Day celebrated the preserving and support of bird life on the private landed estates in Norfolk and the role of their gamekeepers in maintaining the 'balance of nature' by keeping down 'vermin' (the latter included birds like crows and the little owl).

Wentworth Day's moral geography of countryside recreation and his version of sporting naturalism repeatedly returned to the problem of modern leisure and the new natural history within sporting landscapes. In *Broadland Adventure*, his love letter to the Norfolk Broads, 'that enchanting, unspoiled, highly individualistic corner of Eastern England', Wentworth Day railed against the 'cult' of modern bird watching. These birdwatchers, many of whom lacked the countryman's familiarity with wildlife, were an unwanted nuisance, disturbing the birds they had come to see. 'The camera carrying snoopers' of bird 'protectionists' disturbed the more private, exclusive pleasures of Wentworth Day's sporting naturalist with his privileged access to the landscape.

An equal problem was the day trippers and recreationists who swarmed over the Broads in the summer season. With their motor cruisers and loud music, these 'banjo boys', like the birdwatchers, 'vulgarized' the tranquility of the Broads.[20] In a typically misanthropic aside, Wentworth Day noted with pleasure during a cruise down the River Yare on a cold May day that the wintry weather would at least ensure that 'no bright bathing beauties sunning themselves on cabin tops, no banjo-boys in yachting caps and braces, no eager mariners ramming their bowsprits through the porthole unheralded' would disturb his enjoyment of peace and solitude.[21]

Wentworth Day's paean to the landscape of the Broads and the sanctuary that it provided for a truly independent version of the 'English character' was set against another recurring target in his writings: Whitehall bureaucracy and 'socialist planning'. *Broadland Adventure* was published as NC was discussing whether to include the Norfolk Broads in the areas to be designated as national parks. Wentworth Day balked at the prospect of the 'National Parks Administration' getting their hands on the area, fearing they would stifle and control this 'ancient corner of England'. State-sponsored conservation, along with agricultural improvement, had to be resisted in the name of sporting naturalism and the traditions of the private sporting estates.[22]

Wentworth Day's invocation of an older England disappearing under the pressure of commercial development, agricultural improvement, and the unwanted intrusions of 'cockney-shooters', 'banjo-boys', and 'bird lovers' was central to his detailing of the culture of wildfowling and its sporting landscapes. Much of his writing celebrated the long history of wildfowling, particularly in Eastern England, and evoked the 'ghosts of the old gunners'.[23] In summoning these stories of the gunners of old, Wentworth Day documented the impressive number of birds that they routinely killed. He noted, for example, that between 1,200 and 1,400 ducks were killed at 'one of the best wildfowling marshes in England, Old Hall Marshes in Essex, in the 1928/9 season'. Nearly 500 ducks were likewise shot at Cley on the North Norfolk coast between September and February 1930.[24] And comparing his own modest kill of 25 birds shot on an evening shoot, Wentworth Day thrilled his readers with the memory of the 'stupendous bag of 704 brent geese killed in 1860 on the Dengie flats by 32 gunners'.[25]

If a fascination with historical records of killing and the continued precise recording of kills and the size of 'bags' formed an important element within Wentworth Day's understanding of modern fowling, the counting of kills was also allied to a description in his writings of the act of killing. In *Sporting Adventure*, for example, the entry for January captures the moment of the killing of wild geese: 'The whole mud-flat seems to lift in a wave of surging feathers... BOOM... A tongue of flame sears the darkness, an enormous plume of black smoke belches over the water... Geese drop like sacks of wheat, sending up fountains of water. The sky is alive ... with the clamour of fowl... The water is dotted with dead geese, furrowed by wounded striving to escape. The shoulder guns settle them.'[26]

Wentworth Day's recording of this act of killing has a certain restrained, matter-of-fact coolness about it. There is no gratuitous glorification of death and pain. The wounded birds are dispatched humanely. But Wentworth Day is also keen to emphasize across his writings that for the true modern fowler it was not so much the size of the bag that mattered as the quality of the sport. This set the modern fowler apart from the excessive bags and easy shooting of the battue pheasant shooter or the carnage of the grouse moors. Typically, he saw these latter sports as lacking the 'manly' independence and ruggedness of the fowler, who had to use his wits and skills

Figure 5.1 Walter Linnet, wildfowler, and his punt at Bradwell-on-Sea, Essex (1950s).

against that most challenging and wily of foe, the ever-wary flocks of wild geese and ducks.[27]

Wentworth Day was particularly enthusiastic about punt gunning as a form of masculine sporting adventure. This involved the use of a small, low slung, flat-bottomed craft with a huge gun attached. The guns were typically between eight and nine feet long, with a bore of between one and two inches and packing a pound of shot. The fowler lay prone behind the gun, propelling the punt with small paddles and edging along creeks to get close to feeding or resting wildfowl. The punt also held a shoulder gun for finishing off what were called 'cripples' (wounded birds). It was a physically demanding and potentially dangerous sport. Punt guns could explode, killing or maiming the punter, and there was the risk of capsizing if the wind got up and of drifting into deeper wader on the tide. The work of guiding the punt while lying down was both backbreaking and cold.[28] Wentworth Day summarized punt gunning as 'the loneliest, most arduous and most dangerous form of sport left in these islands. [It is] no game for the foolhardy, the weakling or the warm-weather man.'[29]

Alongside the use of punt guns, wildfowlers, Wentworth Day included, practised shore shooting. This required the use of lighter shoulder guns, with the shooter using natural cover, 'digging in' to a shooting pit in the salt marsh, or building a low hide. Wentworth Day detailed the essential attributes of the shore shooter as he awaited the flights of wildfowl: 'Keep down, keep quiet and keep still. Let the birds come to you.' Careful observation of bird behaviour and an understanding of their movements and feeding patterns was crucial to successful shooting, including the positioning of the pit, hide, or place of concealment. Quiet colours and warm clothing were also crucial. These could range from 'your oldest warmest clothes' protected by a 'sober-coloured Mackintosh' to a Norfolk jacket knocked up by your tailor to function as a shooting coat. Rubber thigh boots were also essential, with oilskin trousers necessary when punting. A warm head covering was also important, with either a woollen hat or balaclava recommended.[30]

Another crucial feature of the conduct of the modern fowler concerned the ethics of shooting. Wentworth Day was at pains to point out that the shore shooter needed to work with the limits of his gun and not take long shots or shoot at birds out of range. Long shots meant 'pricked' or wounded birds. 'Wait,' he urged 'til you can see the feathers.'[31]

Shooting birds was further, implicitly, justified by the sportsmen's taking of birds for the pot. Wentworth Day sprinkled his accounts with brief commentaries on how to cook and eat wild geese and ducks. Drawing on his own experience, he offered pithy summaries of what, for example, brent geese tasted like and how to best prepare them: 'Good to eat if you skin and casserole them. They are rich in flavor and taste like a cross between rabbit and pheasant.' Similarly, curlew was good to eat in the autumn and 'properly cooked it is an excellent bird'.[32] Accounts of this sort echoed those of earlier twentieth-century treatises on wildfowling, such as that of J. Nichols, approvingly cited by Wentworth Day in *Broadland Adventure.* Nichols memorably mixed sporting, aesthetic, and culinary assessments of ducks. As he put it: 'The teal is our smallest wild duck, but one of the best for sport; though both for beauty of color and as a delicacy, ... he is surpassed by both wigeon and pintail.'[33]

Wentworth Day's account of the dispositions of the modern fowler was also striking for his emphasis on the joys derived from the landscapes in which wildfowling took place. His writings, as we have already seen, were infused with his strong feelings for the wild places of eastern England and his encounters with birds at dawn and dusk along estuaries, salt marsh, or inland broads. Wentworth Day did much to celebrate the flatlands of Essex – a landscape often marginalized within the dominant landscape aesthetics of the late nineteenth and twentieth centuries. His description of Old Hall Marshes on the River Blackwater in Essex conjured a land of 'creeks, rills, mud flats, saltings, lonely farms and stranded inns which lie in this wilderness of semi-drowned lands [and] forms an oddly out-of-the way corner of twentieth century England'.[34] He eulogized about cloudscapes, of moonlight on dark water, of bright cold mornings. These were the pleasures of the fowler, enchanted by the sights and sounds of wild birds: 'The wedges of wigeon against thunderous skies, the straggling black masses of brent like blown witches over green rollers, the grace and swift elusive beauty of waders. [In these things the fowler] finds joy in a supreme moment in which he can "stand and stare".'[35]

Wentworth Day's writing takes the trouble to notice other, often-rare, birds, and to be gripped by the drama of autumn bird migration and the 'brave [avian] wanderers' arriving, for example, on the North Norfolk coast from Northern Europe. In this way, Wentworth Day presents himself as a careful watcher, sometimes braving the biting cold to observe and not 'merely [there] to shoot wildfowl'.[36] It was a conception of the modern wildfowler as being sensitive to the living world in which his sport was pursued and to natural history and the charm of wild birds.[37]

Wentworth Day's elaboration of sporting naturalism allowed him to inflect his writings in the 1960s towards a natural history rather than an exclusively sporting readership. In *British Birds of the Wild Places,* for example, he reused many passages from his earlier accounts of sporting adventure to describe his encounters with wild birds, but notably shorn of references to shooting.[38] Wentworth Day was also a prescient critic of the use of chemical pesticides in British agriculture, pointing out in an article in *Country Life,* and in a later book, the risks they posed to wildlife and human health.[39] The common ground between natural history and sport carved out by

Wentworth Day found echoes in the early writings of Peter Scott. It formed a point of connection in the creative output of the two men and their shared investment in sporting naturalism.[40]

WILD CHORUS AND MORNING FLIGHTS

Peter Scott (1909–1989) had the mixed blessing of being, as David Attenborough has noted, the son of a national hero: Captain Robert Falcon Scott, 'Scott of the Antarctic'.[41] At the same time, he had the undoubted good fortune of being the son of a formidable and well-connected mother: the society portrait sculptor Kathleen Scott.

Through his mother's social world Scott moved from his childhood years through elite political and artistic circles. Prime ministers, cabinet ministers, and royalty were family friends, with literary figures like Bernard Shaw, Kipling, and H.G. Wells familiar faces at his mother's home in London. Educated at Oundle public school and Trinity College, Cambridge, Peter Scott was an artist, Olympic yachtsman, champion ice skater, glider pilot, wildfowler, and conservationist. He stood (unsuccessfully) as a Conservative Party parliamentary candidate in 1945, the year before he formed the Severn Wildfowl Trust. He was a co-founder of the World Wildlife Fund in 1961, and he became a truly public figure through his role as the principal presenter of the BBC's influential wildlife series *Look* and *Faraway Look* between 1955 and 1970. Later in the 1970s he was associated with Anglia Television's nature documentary series *Survival*, and in 1973 became the first person to be knighted for services to conservation.[42]

With doors opened by his mother, Peter Scott's first career was as an artist, and painting remained an abiding pursuit throughout his life as well as an important source of income as he threw himself into conservation causes. Uninterested in the contemporary European avant-garde in artistic expression, his painting was rooted in the popular romantic tradition. It was his subject matter, though, that came to define Scott as an artist. The subject was wildfowl. Scott had developed a deep interest in natural history as a child, but it was during his years as a Cambridge undergraduate that he discovered the lure of wild ducks and geese through the sport of wildfowling.

Borrowing his late father's old gun, Scott threw himself into the world of wildfowling.[43] Scouring wildfowling books, he learnt that the North Norfolk coast around Wells-next-the-Sea was 'the mecca'.

During the Christmas vacation of 1927 he travelled there to experience wildfowling first hand. Scott later recalled the electric thrill of hearing and seeing large flocks of pink-footed geese out on the marshes, both at night and at dawn. It was a sound that henceforth never failed to excite and stimulate him. As he put it: 'The thrill which I felt on that first morning is repeated every time I see the great skeins stretched across the sky. The spell is as strong as ever.'[44]

While the trip to North Norfolk provided Scott with memorable encounters with wild geese, he failed to shoot any. Driven almost to the point of obsession, he spent many days and (particularly) nights slipping out of his college rooms to pursue geese in the wetlands and fens of nearby Cambridgeshire. With a small inheritance at his disposal, Scott designed and had built a punt, which he named *Grey Goose*, and acquired an old punt gun and other wildfowling kit to furnish his ambition to stalk and shoot wild geese.[45]

Embracing the pursuit's dress codes, Scott styled himself as a modern fowler. In a self-portrait from 1935 that formed the frontispiece for his book *Morning Flight*, he depicted himself in roll-neck jumper, earth-toned wool sports jacket, and thigh boots, shotgun across his arm, a steely look in his eyes, and the dark skies and flat looming salt marsh behind him. Similarly, a photographic portrait of him taken around 1931 shows Scott standing proudly next to his punt with the flat mudflats behind him. In thigh waders, oilskin skirt, cap, and jumper, he looked the part of the young modern fowler. Tellingly, the reproduction of the photograph in his autobiography appeared opposite another image of Scott from 1931, this time embodied as the sleek, balletic ice skater clad in figure-hugging black top and tights. It revealed another side of Scott and a different self-styling through the codes of a different modern winter sport. Here were two of the multiple identities or personas that Scott was able to inhabit as this young man of action pursued different kinds of adventurous recreation.

In the accounts that Scott gave of his goose hunts – recorded first in his wildfowling diaries and later reworked in two books he published in the 1930s, *Wild Chorus* and *Morning Flight*, and recapitulated in his autobiography (published in 1961) – he conveyed the extensive planning, adventure, physical ordeal, and disappointments integral to wildfowling. A punt gunning trip, in particular, was like a 'campaign': 'It requires organization, generalship and seamanship. It is difficult and arduous – often disappointing and sometimes a little dangerous. In this sophisticated land of ours it is one of the few remaining sports that offer adventure.'[46]

Punt gunning took Scott into wild and dramatic landscapes, particularly the marshes and creeks of The Wash, and he was moved by the bleakness and loneliness of these places, particularly in the pre-dawn grey of a winter's morning. As he put it in *Morning Flight*: 'If you go after wild geese you will assuredly go into beautiful wild places at the most beautiful times of the day – at dawn, dusk, or moonrise – and best of all you will hear them [the wild geese] call.'[47] In other accounts he recalled that failing to shoot birds was more often than not made up for by the spectacle of wild birds. As he put it, 'our blank day was worthwhile if only for the memorable sight of 3000 geese "whiffling" down onto the mud'.[48]

The sound of geese was as central to Scott's experience of wildfowling as were the chase and the visual spectacle. His accounts presented a sonic geography of the places of encounter with wild birds, and it was the calls of wild geese that helped to fix and centre him in these landscapes. Referring to goose hunts in the East of England, Scotland, and Ireland – as well as forays further afield in Hungary, Persia, and Canada – Scott suggested that 'the music of the great skeins is always moving in its grandeur, always perfectly appropriate to the wild places in which it is heard'. Elsewhere he confessed: 'I hear the geese calling – a music of indescribable beauty and wildness.'[49]

If the soundscape of the marshes and the beauty of the landscape were central to his experience of fowling, so too were the hunt and the chase. Stalking wild geese and ducks was a battle of wits with wary and watchful birds, and managing to get close to these wild creatures through the careful manoeuvring of a punt generated a particular intensity of feeling: 'As the punt glides closer the excitement grows, until one feels as if one had just run 100 yards all out.' The wildness of the birds was tantalizing, and getting close to species that were both wary of people through centuries of being hunted and had the wind of far off places like the Arctic tundra in their wings added to their romantic allure.[50]

If the wildness of the birds and the intensity of the stalk created the adrenalin rush of hunting, these emotions were released or consummated with the firing of the gun. Recording a trip to Hungary in March 1936, Scott noted, 'the first evening ... I opened with a nice right and left and later I got two birds'. During the week he collected a bag of 350 birds.[51] An earlier expedition in 1929 saw Scott successfully stalk much-sought-after brent geese: 'The trigger ... was firmly gripped and pulled – then came the thunderous roar and a cloud of smoke, and from the side of the cloud flew 3 brent geese. When the smoke [had] cleared 4 lay dead upon the water.'[52] A Christmas punting trip also proved successful, with Scott documenting the birds killed – some from the punt gun and others, wounded birds, finished off with the shoulder gun: 'After half an hour of strenuous running about and shooting ... we picked up 31 birds. 27 were pintails, 3 wigeon and one an unlucky shelduck which had been amongst them.'[53]

Scott was relaxed about his power of life and death that he had over these birds – creatures that he evidently also found attractive and beautiful. In his autobiography he justified his shooting as the 'primitive enjoyment of the hunter ... part of man's instinct to hunt'. Geese were traditional human quarry: they had been shaped as a prey species by being hunted, and hunter and goose were locked into an age-old relationship that could legitimately be concluded with the death of the birds.[54]

Scott's belief in the rightness of shooting was also combined with a passion for seeing rare and charismatic species of geese. His trip to Hungary in 1936 had been largely motivated by a deep desire to see the red-breasted geese that wintered on the Hungarian plains. This passion for finding and seeing rare geese gave Scott immense pleasure. When, just after the War, he was able to see only the second record of a lesser white-fronted goose in Britain, he was thrilled. The excitement of finding a rare bird prompted in him a different emotional uplift from shooting birds and he tellingly compared it to the physiological effects of a fine piece of classical music: 'In the instant my binoculars hit upon it, I realized that it was a lesser white-front. My spine tingled delightfully as it does in the slow movement of Sibelius's violin concerto.'[55] Here was a patrician view of rarity-chasing birdwatching informed by Scott's refined cultural formation.

Scott's painting sought to capture the intense excitement and drama of wildfowling – as well as, sometimes, rare geese – though his art was notably shorn of the depiction of acts of killing. Unlike the tradition of still life oil painting in which game birds often figured as decorative, if lifeless, bodies in the assemblage, or the tradition of bird illustration that used museum specimens as the basis for wildfowl portraits, Scott's artwork was of living birds in their natural environments.[56] In this regard, his paintings were also set apart from the photographic conventions of the sporting estates and sporting press, in which hunters would pose next to a mass of successfully killed birds or be depicted holding dead birds or with their trophies lying nearby.[57] If Scott's paintings captured the drama of birds as encountered through wildfowling, the act of shooting was not represented and nor were birds seen dropping from the sky. What he repeatedly depicted were ducks and geese in dramatic weather conditions, flying in snowstorms, against thunderous skies or on bright clear mornings. The patterns of clouds, the chiaroscuro drama of dusk or moonlight flights, and the pale light of dawn all recurred in what were effectively landscape paintings with birds sometimes reduced to darks skeins across the skyline. At other times, Scott depicted close-up views of birds, typically in flight, springing from pools, or landing in frosty fields. These contrasting perspectives – from distant flocks against big landscapes to close-ups of intense action – codified the stereoscopic perspective of the wildfowler drawn from Scott's own experiences. Critics praised the 'feeling of life' in his paintings, the sense of wildness in them, the way his 'brush seemed charged with fresh air and salt breezes', and his capacity to capture in oils the character of birds.[58]

Scott's style and subject matter proved lucrative. By 1939 he was making around £4,000 a year from his artwork. Central to Scott's commercial success was his ability to exploit his paintings for popular markets. In 1930 one of his paintings, 'Taking to wing', was reproduced in the Medici Society's 'modern artists' series, with the picture being made available as both large- and small-format prints, as a greetings card, as a Christmas card, and as table mats. Between 1936 and 1982 Ackermann's Galleries in New Bond Street, London – where Scott, through his mother's introduction, held annual exhibitions between 1936 and 1940 – made prints of twenty-nine of Scott's paintings.

And there were other commercial sidelines, too: Player's cigarettes paid Scott £1,000 for a set of twenty-five small wildfowl paintings to use on cigarette cards; and the pottery manufacturer, Wedgewood, commissioned a painting for a set of plates.[59] Scott's art became a recognizable and popular style of wildfowl painting, reaching beyond the immediate audience of wildfowlers and those interested in natural history and connecting with a wider public.

Scott's painting and his writing about wildfowl shared much with James Wentworth Day's commentaries on the sporting landscapes in which he pursued wildfowling. Both placed the encounters with wild ducks and geese within a deep feeling for and awareness of British landscapes, particularly the lonely places of salt marsh and estuary. Both were moved by the changing patterns of sky and clouds, and by the emotional geography of wild shooting within sporting landscapes. Peter Scott, however, unlike Wentworth Day, began to develop doubts about the morality of shooting. Once a committed fowler, his reservations hardened following one incident in particular. During a goose shoot an adult bird was wounded and crash-landed in an inaccessible part of the marsh. The bird could not be recovered and was still there the following day, its partner flying around it calling. Scott was deeply troubled by the incident, questioning his right to inflict such suffering on a wild creature. As he recalled, once the doubts and disquiet took hold of him he could no longer enjoy shooting: 'So I sold my guns and I no longer shoot.'[60]

Scott's internal conflict about the morality of shooting sat alongside a developing interest in collecting ducks and geese. At the old lighthouse that he had bought on the Nene Washes in the 1930s, he built up a collection of formerly wild geese and new captive-bred birds. His collection became an increasingly important source of pleasure, with Scott depicted in photographs from the time as goose farmer and aviculturalist.

It also shaped a new kind of intense emotional relationship with one wild goose in particular. In the autumn of 1936, a young pink-footed goose that had become separated from its family group on migration was attracted to Scott's pinioned collection of birds. It flew in and joined the captive flock. The bird immediately captivated Scott. He called her 'Annabel', 'a lost child', and delighted in his close contact with her. When she returned the following autumn, he was thrilled. As the bird approached Scott and his feeding bucket, he was nearly overwhelmed at her return: 'There she stood, a plump little round person, with her queer angular forehead, her unusually pink bill pattern. To me she was as recognizable as a stray sheep to a shepherd.'[61] In individualizing the bird and giving it human attributes, Scott was moving away from the dominant view of geese within wildfowling as an undifferentiated mass of birds and projecting intense familial emotions onto the goose. While James

Wentworth Day, like many wildfowlers, was close to his gun dogs (including the improbably named Mr Soapey Sponge), with the canines constituting important companion animals for him, Scott's emotional attachment to Annabel represented a different ordering of human–avian relations from Wentworth Day's. When Annabel failed to return in the autumn of 1938, Scott was bereft, grimly conceding that the greatest threat to wild geese once they had fledged came from the wildfowler's gun.[62] His relationship with Annabel, however, would be repeated with other birds and was expressed in a more scientific form in his later studies of wild swans, where he developed a system of identifying (and naming) individual birds that became familiar companions around his home in Gloucestershire.[63]

Scott's interest in building a collection of captive wildfowl matured as an idea over the years, and on his return from active service with the navy during World War II he set about putting his thoughts into action. The plan was for some kind of expanded collection where wild and captive geese could be shown to the public, and where protection, scientific study, and education could be combined.

Visiting an area of marsh and grassland in Gloucestershire to witness its large flocks of wintering geese, Scott was struck by how perfect the site would be for his new wildfowl refuge and collection. The area was in the Berkeley Vale in Gloucestershire. He had already visited it in the late 1920s because of its reputation as a good shooting area. The Berkeley family held shooting rights over the land and, with the landlord, Scott sought out the geese on an area known as The Dumbles. Approaching the birds unseen along the raised riverbank, Scott crept into one of the wartime pillboxes that had been built on the marsh to get close to the geese. Peering through the opening of the pillbox, he was delighted to see the flock of geese very close by. The encounter with the geese convinced Scott that the Vale of Berkeley, close to the village of Slimbridge, was the ideal place in which to establish a new sanctuary for wild geese.

His own viewing of the geese from the pillbox became the prototype in his mind for creating a new kind of experience of wildfowl for the general public. Observation posts or huts would give those who had never seen wild geese up close the opportunity of 'watching the wildest of wild birds at really close range'.[64] To the close encounters with wild geese would be added the attraction of a captive collection, with the refuge being used to both study geese and to educate the public about them and the wider cause of wildfowl conservation.

In November 1946, at a meeting held at the Patch Bridge guesthouse in Slimbridge, the Severn Wildfowl Trust was established as the vehicle for his new project.[65] He, and it, would play a significant role in the post-war debates about wildfowl, wildfowling, and bird protection.

PROTECTING WILD DUCKS AND GEESE

The idea of creating nature reserves, bird sanctuaries, and wildfowl refuges had grown, as we saw in chapter 3, as a dimension of post-war planning during World War II. The programme for establishing both local and national nature reserves was given legal force with the passing of the National Parks and Access to the Countryside legislation and the creation of Nature Conservancy in 1949. Plans for new protected areas, including those around the coast and at wetland sites, formed part of this plan.

While conservation groups saw this as the culmination of a long-standing campaign for greater protection of Britain's 'wild heritage', wildfowling groups and those at the upper echelons of WAGBI were more troubled by the proposals. Organized wildfowlers certainly felt that new refuges and reserves were yet another mechanism for limiting and restricting their sport. Since the late nineteenth century, a series of parliamentary bills had placed constraints on wildfowling. The Wild Birds Protection Acts (1880 and 1881) had entrenched a close season for wildfowl, running from 1 March to 1 August, in order to protect breeding bird stocks in the face of evidence of big declines in the numbers of wildfowl.[66] Further incremental restrictions on the sport were implemented in the Wild Birds Protection Bill (1925) and the Wild Birds (Ducks and Geese) Protection Bill (1939). The latter legislation extended the close season from 1 February to 11 August, adding all wild geese to the schedule and banning the importing of wild geese and wild ducks during the same period.[67]

The British section of the International Committee for Bird Preservation (ICBP) was central to the promotion of the 1939 Bill. The ICBP, with its strong representation of leading American bird protectionists, had earlier sought to encourage international cooperation to protect stocks of migratory geese and ducks in Europe and North America in the wake of growing concerns about big declines in their populations, particularly in the United States and

Canada. The ICBP had established a subcommittee in 1916 to study the distribution and populations of wildfowl. This became the International Wildfowl Bureau, eventually coordinating wildfowl counts across twenty-one countries.[68]

In the 1920s and 1930s, WAGBI, along with the sporting press, had been supportive of these moves to protect wild ducks and geese, in large part to ensure the safeguarding of stocks available to wildfowlers. Both for protectionists like the ICBP and for sporting interests, the introduction in the United States and Canada of tighter controls on bag size and the creation of breeding and resting sanctuaries for wildfowl was seen as an exemplary way to proceed, having helped to add appreciably to the numbers of wildfowl. By the late 1940s, however, WAGBI had become more concerned about the threat to free shooting from the establishment of conservation areas in the United Kingdom.[69] In this context, the formation of the Severn Wildfowl Trust and the reserving of the New Grounds at Slimbridge was a source of concern.

Peter Scott's programme for Slimbridge combined three distinct approaches to what he saw as the 'scientific study' of wildfowl. These were the study of the flocks of wild geese; the catching and ringing of ducks in the old duck decoy at the New Grounds; and the assembling of a comparative international collection of wildfowl for study and education purposes.[70] These scientific studies would provide new knowledge to assist in the protection of the world's wildfowl and help in arresting their decline.

Scott was careful, however, to present these studies as being of relevance to all those groups interested in wildfowl.[71] Reaching out to sporting interests was particularly important. Scott combined this with his characteristic ability to draw on his connections with the great and the good to ensure wide support for the Severn Wildfowl Trust. Field Marshall Lord Alanbrooke, former chief of the Imperial General Staff and a keen sportsman, agreed to become president of the Trust, and he in turn suggested that Sir Archibald Jamieson, chairman of the manufacturers Vickers and also a 'shooting man', become treasurer. Max Nicholson, the director-general of NC, was soon recruited as one of three vice-presidents, along with the 'new ornithologist' James Fisher and the ethologists Niko Tinbergen and Konrad Lorenz.[72] More telling, however, were the appointments of Dr Jeffery Harrison, a high-profile member of WAGBI, to the Trust's scientific advisory committee and of the Duke of Beaufort, president of the British Field Sports Society, as one of the trustees of the Trust.[73]

During the early years of the Trust's existence, Scott continued to occasionally shoot, and he was keen to present the Wildfowl Trust (as it was renamed in 1954) as a friend and not a foe of wildfowlers. The Trust took over the wildfowl counts initiated by the ICBP and enlisted the support of fowlers and local naturalists in its annual winter surveys. Scott also spelled out in the sporting press that the Trust's activities were not set against wildfowling or wildflowers. Wildfowl, he conceded, were a natural resource that could and should be harvested, with such harvesting benefitting from population studies of wildfowl. Likewise, 'a moderate and sensible policy for sanctuaries arranged and planned with the full cooperation of the local wildfowlers could work in the interests of both conservationist and sport'.[74]

Scott, however, was a controversial figure for many wildfowlers and for the sporting press in the early 1950s. Their antipathy towards him was intensified in the debates surrounding the passage of the Protection of Birds Act in Parliament in 1953 and 1954.

'A WILDFOWLER'S CHARTER' OR THE WORK OF RABID PROTECTIONISTS?

When the Protection of Birds Bill was published in November 1953, *Shooting Times* was generally supportive. The consolidation and simplification of the law was welcomed and an editorial in the paper praised the proposed outlawing of shooting in August. As the Bill progressed through parliament, however, WAGBI and *Shooting Times* became more concerned. The decision of Lady Tweedsmuir, the Bill's sponsor, to go against the recommendation of the Home Office's Wild Birds Advisory Committee to extend the wildfowling season until late February caused great anger. Editorials in *Shooting Times* saw the Bill as a threat to free shooting on the foreshore. The paper detected the influence of birdwatching societies in this change to the Home Office committee's proposals, seeing many birdwatching clubs as 'almost fanatical in their animosity to the shore shooter'.[75]

A review of the Bill in *Shooting Times* in January 1954 found other evidence of the influence of the 'bird protection societies' on legislators, seeing in the provision of new statutory powers for the creation of bird sanctuaries a sign of their power and influence. These new proposed powers given to the Secretary of State provided an opportunity for conservationists to 'seize great stretches of foreshore … until they have driven [out] the wildfowler'.[76] Tower-Bird, the journalistic pseudonym of Noel (Tim) Sedgwick, underlined his concerns about this dimension of the Bill, fuming, in terms reminiscent of Wentworth Day, about the alliance of 'rabid protectionists' and 'planners' who were working, in an 'extension of the Welfare State', to drive state-sponsored sanctuaries over the established rights of private sporting interests.[77] The danger facing birdlife, he concluded, was not from the private estates that were 'the true bird sanctuaries of the British isles', but from the roving bands of organized '"townee" birdwatchers with their little notebooks' and cameras who disturbed rare and nesting birds and showed a 'disregard for private rights'.[78]

As *Shooting Times* continued its sniping at the Bill, another flashpoint occurred in relation to the protection given to quarry species, especially the brent goose. The decision of Lady Tweedsmuir to amend the Bill at the committee stage to give brent geese complete protection, for an experimental period, incensed *Shooting Times*. They saw Peter Scott, and the Severn Wildfowl Trust, as the source of this policy, and their accusation was not without foundation.

In defending her decision, Lady Tweedsmuir drew on evidence given to her by Max Nicholson about the large flocks of brent geese that had formerly been seen both in Britain and across Northern Europe. The birds had undergone a sharp decline, and the arrival of 'great armies' of the birds noted by wildfowler–naturalist Abel Chapman at the turn of the century had become a thing of the past. Tweedsmuir suggested that the failure of eelgrass, the principal food of the birds in winter, was behind these declines, but that shooting exacerbated the problem by preventing the birds from recovering their numbers. Evidence supplied by Peter Scott of the toll taken in particular by punt gunners on the birds in the hard winter of 1946/47 gave an emotional weight to the claims. As Lady Tweedsmuir recounted it, forty-six punts working in one Essex estuary had killed around 150 brent geese. Conceding that her husband was a 'great wildfowler' and that she enjoyed accompanying him in his pursuit, she nonetheless concluded that the birds needed to be protected in order to recover their population. As one of the MPs present later recalled, Lady Tweedsmuir's speech on the need to protect brent geese was the 'most dramatic moment' of the committee stage of the Bill.[79]

The letters pages of *Shooting Times* bristled with hostility towards the Severn Wildfowl Trust, with correspondents seeing Scott's figures about the killing of brent geese as exaggerated and unsubstantiated.[80] The feverish concerns of wildfowlers about the influence of figures like Scott and of NC over the passage of the Bill was not entirely borne out by the parliamentary debates. These revealed strong support from a group of MPs and members of the Lords, often self-confessed wildfowlers, for regulated wildfowling. Encouraged by the active lobbying of WAGBI, parliamentarians like Colonel Clark (Conservative MP for East Grinstead), Sir V. Raikes (Conservative MP for Gartson), and B. Burden (Conservative MP for Gillingham) were successful in persuading Lady Tweedsmuir to change her mind over the end date of the close season.[81] In a significant revision of the Bill, Tweedsmuir announced in the House of Lords in March 1954 that she wished to move an amendment to enable the shooting season along the Crown foreshore to be extended until 20 February. This she felt was the 'fairest compromise between sporting and protectionist interests'.[82] Now it was the conservationists who were angered by the shape of the Bill. In a letter to Max Nicholson, Peter Scott expressed his exasperation: 'I think it is quite distressingly weak-kneed of Priscilla Tweedsmuir to yield to the February 21st issue... Many species of wildfowl would be better off if the Bill were talked out... *The Shooting Times* is more than usually vitriolic about us both this week.'[83]

Max Nicholson shared the frustration of his friend. With the Bill completing its passage through the House of Commons that April, Nicholson confessed to a colleague that he was greatly disappointed about the concessions given to wildfowlers. As he put it: 'As Lady Tweedsmuir herself frankly admitted ... the House has somehow persuaded itself that this virtual capitulation ... to the small organized minority in the face of the facts represents common-sense whereas in our view it is nothing but a minor Munich.'[84]

If Nicholson felt the Bill appeased the unwarranted demands of wildfowlers, the final shape of the legislation similarly troubled pro-conservationist MPs in Parliament. A number, including Frank Hayman (Labour MP for Falmouth and Cambourne), dubbed it 'a wildfowler's charter'.[85] The key schedules of the Bill as they bore upon wildfowling certainly represented a partial victory for wildfowlers. The new dates of the open season for foreshore shooting, running until 20 February, represented a three-week extension of the free shooting season. Another aspect of the new Bill – the restriction placed upon the muzzle diameter of punt guns that required them to be no more than 1.75 inches in size – had little material effect since the few remaining punt guns in use usually had muzzles of no more than 1.5 inches. Furthermore, no new punt guns were being manufactured at the time and the sport was already in sharp decline.

Brent geese, however, had been given year-round protection, and the ban on shooting barnacle geese was also maintained.[86] Tower-Bird, in a *Shooting Times* article, saw the battle over the Bill as having been, in the end, 'a great victory for wildfowlers'.[87] Looked at in another way, however, the parliamentary and extra-parliamentary battles over the Bill also revealed the wide area of agreement that existed between most wildfowlers and conservationists, regardless of the animosity that had been aroused between them. WAGBI and its allies supported controlled and regulated shooting and the need to protect wild populations of ducks and geese. Many wildfowlers and MPs also confessed to their combined identification as both keen naturalists and keen sportsmen.

The desire to build on and consolidate this common ground, in large part to sustain the momentum of the new Bill and to ensure wildfowlers supported its application, lay behind Max Nicholson's idea to initiate his 'tea parties' at NC headquarters. As the informal gatherings developed at NC, representatives of WAGBI also met with key figures in the RSPB to identify common concerns and areas where they could support each other.[88]

What was telling about these meetings between the upper echelons of organized wildfowling and conservation was the degree of crossover in interest and membership between the two sides. Geoffrey Dent, a leading figure in the RSPB present at the meetings with WAGBI in 1957, was a member of the British Field Sports Society (and a Master of Hounds). The chairman of NC was a prominent wildfowler, and its director in Scotland was the president of the Tay Valley Wildfowlers Association.[89] These crosscutting affiliations expanded as good relations between WAGBI, the RSPB, NC, and the Wildfowl Trust developed through the 1960s. By the time the second edition of *The New Wildfowler*, a book sponsored by WAGBI, was published in 1970, both the RSPB's secretary Peter Conder and his predecessor Philip Brown were members of WAGBI, together with the Society's Scottish representative, George Waterston.[90] Max Nicholson, former director-general of NC, had also joined. From the other side, Jeffery Harrison, a member of WAGBI's scientific advisory committee, who had been a member of the Wildfowl Trust's advisory committee for many years, had become a member of the RSPB Council. John Anderton, WAGBI's honorary secretary, was a self-declared RSPB member. In a valedictory pamphlet published in 1970, WAGBI, NC, and the Wildfowl Trust celebrated the successful collaborations between fowlers and conservationists that had developed since the divisions over the Protection of Birds Bill, marking what the pamphlet called the success story of this 'triumvirate'.[91]

At the heart of the building of this alliance between wildfowlers and conservationists was the self-consciously forward-looking idea of the 'new wildfowler'. As wildfowling grew in popularity through the late 1950s and 1960s, with WAGBI's membership rising from less than 500 in 1957 to more than 23,500 by 1972, pressure on the countryside grew from these new wildfowlers and from wider members of the public.

WAGBI and the sporting press sought to elaborate on the ethos and privileges of the 'new wildfowler'. Central to this was a modernization of the idea of wildfowling as a pro-conservation activity, with WAGBI casting itself as an organization representing 'wildfowler conservationists'.[92] Supporting wildfowl refuges and the creation of NNRS on traditional wildfowling sites was central to this approach. It went along with the winning of privileged access for WAGBI members and other 'legitimate sportsmen' to permit-based shooting on these reserves.[93] As an editorial in *Shooting Times* put it, 'a suitably administered series of wildfowl refuges should be regarded as a natural counterpart to sport'.[94]

Excluded from what one correspondent to *Shooting Times* described as the 'closed-shop' of local wildfowling clubs and their access to reserves were those against which the 'new wildfowler' was defined: the 'illegitimate shooters', the 'hooligans', the 'marsh cowboys', and the 'indiscriminate shooters'. *Shooting Times* and books such as *The New Wildfowler* demarcated the restrained and controlled dispositions of the modern sportsman from these unwanted social types. In his essay on conservation and bird protection in the WAGBI-sponsored book, Philip Brown called the unwanted shooters 'inexperienced louts'.[95] Peter Scott had earlier, in the 1961 edition of *The New Wildfowler*, castigated them as 'hooligans'. The editors of *The New Wildfowler* also warned against the problem of 'marsh cowboys' and the damage done to the cause of wildfowling.[96] In an article in the RSPB's members' magazine, John Anderton, honorary secretary of WAGBI, railed against the 'unsporting and irresponsible behavior of gun toting youth – and sometimes the not so young – known as ... marsh cowboys'. 'They are a curse,' he concluded to 'both the naturalist and the bona fide shooting man.'[97] In a similar vein, a comment piece by Peter Conder, the RSPB's secretary, praised the work of WAGBI in improving and enforcing an ethics of shooting and its 'strong and oft repeated condemnation of marsh cowboys'.[98] An agreement reached between the Norfolk Naturalists Trust and the Great Yarmouth and District Wildfowlers Association over shooting rights on the new bird sanctuary similarly spoke of the need to ensure that 'all but *authorized* shots should be curbed in the area'.[99]

The alliance formed between sporting and conservation interests was also underpinned by the two groups' shared identification of growing threats posed to the birdlife and the environments that they wished to protect. Philip Brown spelled out the 'necessity for all concerned to unite in striving to preserve essential ... habitats, for without these there will be no ducks, no geese, no birds, no bird watching, no photography, no "fowling": we shall have oil-refineries, holiday camps, nuclear power-stations and caravans'.[100] An editorial in *Shooting Times* made a similar point, noting the common cause of 'good sportsmen' and 'good protectionists' in the face of 'economic pressures on land usage, coupled with ever-increasing disturbance due to recreational activities'.[101] WAGBI's John Anderton, in an article in *Birds*, laid out the threats from the drainage of marshland and recreational pressure upon wildfowl habitats, emphasizing the importance of cooperation between organized wildflowers and conservationists in the creation of a national network of wildfowl refuges.[102]

This 'common cause' was mobilized in the campaign led by the RSPB to protect the area of Maplin Sands and Foulness from being the site of the proposed new London airport in the early 1970s. WAGBI and wildfowlers in Essex and Kent were among thirteen organizations – including the RSPB, the Wildfowl Trust, the ICBP, the BTO, and the Essex Birdwatching Society (EBWS) – that supplied joint evidence for the public inquiry in stage two of the Roskill Commission on the siting of London's third airport.[103]

Shooting Times, under Philip Brown's editorship, deepened the links between wildfowling, bird conservation, and birdwatching. Jeffrey Harrison, one of the clearest embodiments of the wildfowler–conservationist, produced articles detailing his birdwatching trips in which he emphasized the observation and photographing of birds.[104]

Among the forging of common ground and interests between the 'new wildfowlers' and the bird conservationists, however, differences did remain. In his contribution to *The New Wildfowler*, Peter Conder, having justified wildfowling as the cropping of a natural surplus and as being profoundly allied with conservation, noted that wildfowlers shared birdwatchers' liking for 'wild, wet open spaces'. But he also conceded that there was a fundamental difference between the two pursuits. As he put it: 'When a mallard comes into range the bird watcher with his binoculars looks at it, admires its plumage and tries to understand its behavior. Whereas the wildfowler shoots it and enjoys its flesh.'[105]

This fundamental difference in the ordering of human–avian relations was picked up on by RSPB members in the late 1960s and early 1970s, with a sustained correspondence taking place in the Society's members' magazine about the ethics of wildfowling. The flurry of letters was prompted not only by the public statements of leading figures within the RSPB, such as Peter Conder, but also more immediately by the publication of an article in *Birds* titled 'The wildfowler as conservationist' by Jeffery Harrison. Prefaced by an editorial praising WAGBI and its pro-conservation stance, Harrison saw the debates on the Protection of Birds Act as having been generative in helping wildfowlers, and birdwatchers, to shift their respective outlooks. From the wildfowling side, the educational work involved in creating a 'new wildfowler' had not only encouraged fowlers to see the importance of working to protect suitable wetlands, but had also prompted them to create new wildfowl reserves and support duck and goose breeding schemes to increase the stocks of wildfowl.[106] Harrison himself had pioneered the creation of a new wetlands reserve for wildfowl, eventually working with the Wildfowl Trust on a joint venture to develop an old gravel pit in Sevenoaks in Kent. He was also a leading player in the WAGBI-sponsored duck breeding scheme and in the introduction of Greylag geese into southern England.

Not all readers of the RSPB's magazine, however, were persuaded by this cosy relationship with wildfowlers or by the idea of the wildfowler as conservationist. Brian Mitchell and Brian Hawkes, two Kent birdwatchers, balked at the suggestion that bird conservation and wildfowling could go hand in hand. 'Many conservationists,' they suggested, 'most strongly deny the morality, and therefore the eventual social effects, of the behavior and attitudes of people whose pleasure involves killing.' They went on: 'The sports of wildfowling and game-bird shooting enact quite calculated and self-indulgent violence and objections to these sports are therefore radical and important: the conservation cause can only be forwarded by stating them… Wildfowlers must know that there are very many serious and genuine opponents to their behavior and philosophy.'[107]

Others supported Mitchell and Hawkes's letter. D.S. Bunn, for example, thought that the fact that wildfowlers contributed towards conserving their quarry was irrelevant. The ethics of the matter, he contended, 'lie in whether a species should be conserved so that pleasure can be gained from hunting or shooting it'.[108] Defenders of wildfowling, however, also took to the letters pages of *Birds* to support the practice. Richard Moles, a member of both the RSPB and WAGBI, conceded that not all wildfowlers were, as yet, conservationists. 'Some,' he noted, 'may claim to endorse the spirit of the "New Wildfowler", yet a walk along duck frequented foreshore … will show that the "marsh cowboy' is most certainly not extinct.' However, the wildfowlers associations, he suggested, were working hard to encourage a new ethics of shooting and the creation of new habitats for wildfowl.[109] S. Cowdry was more robust in his response, reminding readers of *Birds* that 'responsible shooting is legal [and that] man is a hunter by nature and bird watching is only a form of this hunt'.

Unsurprisingly, the Rt Honorable The Earl of Mansfield, president of WAGBI, also wrote to defend both his association and wildfowling in general. He emphasized the close collaboration between WAGBI and the Wildfowl Trust, the RSPB, and NC, particularly over the protection of wetlands and the creation of refuges. He also quoted Peter Scott's defence of controlled wildfowling from *The New Wildfowler*, where Scott had justified it as the harvesting of ducks and geese as a natural resource.[110]

These defences of wildfowling stirred Mitchell and Hawkes to return to the letters pages, however. Upping the ante, they reiterated their claim about wildfowling being a 'deliberate, self-indulgent [act of] killing'. In emotive terms, they reminded wildfowlers, and the readers of *Birds*, that wildfowling was 'the deliberate act of smashing a living creature's body with hot metal'.[111]

CONCLUSION

The growing ethical distance between wildfowlers and at least a section of the RSPB's membership began to fracture the relationship between these two bird-centred cultures of nature by the early 1970s. It was the product of the changing social make-up of the Society's membership. As more urban and suburban birdwatchers joined the RSPB, cultural differences between them and the elite leadership of the Society disturbed the close social connections that existed at the upper echelons of bird protection and wildfowling.

This was not the only source of the opening up of divisions, either. For high-profile figures like James Wentworth Day, the ongoing protection given to some quarry species needed to be challenged. In an article in *The Times* in March 1974, Wentworth Day pressed for the total protection of brent geese to be lifted. They were now so common, he suggested, that they had become a 'menace' to coastal farmers. It was time for farmers to be allowed to shoot the birds and protect their crops.[112]

Wentworth Day's article provoked a response from three of the leading figures that had helped to forge the post-war alliance between conservation and wildfowling: Peter Conder, Peter Scott, and John Anderton (WAGBI). Challenging Wentworth Day's claims and insisting on the importance of the brent goose ban, they again articulated the common ground between the two pastimes of birdwatching and wildfowling. They noted the growing pressure from development on the estuarine habitat favoured by the birds, casting the brent goose not as a 'national menace' but as a 'national asset'.[113]

If the alliance of wildfowling and bird conservationism was being tested by figures like Wentworth Day as much as by the debate from birdwatchers about the morality of killing, the growing segregation of the conservation and wildfowling landscapes was also having a profound effect. For example, shooting at Hickling Broad in Norfolk – designated a national nature reserve (NNR) in 1958 – was ended in 1965. A new conservation 'water trail' for visitors was established in 1970 in the area that, as David Matless notes, had only recently been vacated by fowlers.[114] At the Norfolk Naturalists

Trust's reserve at Cley Marshes in North Norfolk, acquired in 1926, where conservation had previously coexisted with the winter duck shoot, wildfowling also ended in 1966 when the site was declared a sanctuary under the Protection of Birds Act legislation. Similarly, the shooting rights held by the key landowner, Stuart Ogilvie, on the RSPB's Minsmere reserve in Suffolk came to an end when the Society bought the land outright in 1977.

During this same period, the Wildfowl Trust's reserve at Slimbridge, where there had been no shooting since the late 1940s, was a popular visitor attraction. It drew in a large public, with its average annual gate between 1968 and 1971 being around 200,000 paying visitors, excluding the Trust's 9,000 members.[115] Through its chain of nineteen observation huts and its large collection of captive-bred and pinioned wildfowl from around the world, Slimbridge offered a recreational orientation to ducks and geese rooted in the close observation of wild birds and a domestic proximity to the captive ones. In an area originally called 'Rushy Pen', wild and captive birds were fed together. The feeding of the birds became a major part of the visitor attraction at Slimbridge, with Rushy Pen renamed Swan Lake in 1965. When a similar arrangement was established at one of the Wildfowl Trust's other reserves at the Ouse Washes in Cambridgeshire, with floodlit feeding of the birds taking place for visitors in front of a heated bird hide, one observer dubbed it a 'swan et lumiere' experience.[116]

Scott himself used the favourable conditions for viewing wild swans on Swan Lake at Slimbridge to begin his study of the bill patterns of the wild birds in order to identify and study nearly 200 individually recognizable birds (see figure 5.2). This combination of intimate study of birds seen as individuals and the theatre of birdwatching was a long way from Scott's youthful participation in the sporting adventure of wildfowling. It represented a deep reordering of the practices, technologies, and motivations for encountering wild ducks and geese.

The tensions, divisions, and differences between wildfowlers and birdwatcher–protectionists through much of the period from the 1920s to the 1970s tended to overshadow the commonalities that linked the two bird-centred cultures of nature. Both pursuits shared a love of the landscapes and places in which wild ducks and geese were encountered, and the beauty, spectacle, sound, and romantic associations of wildfowl riveted both. At a national level, the groups

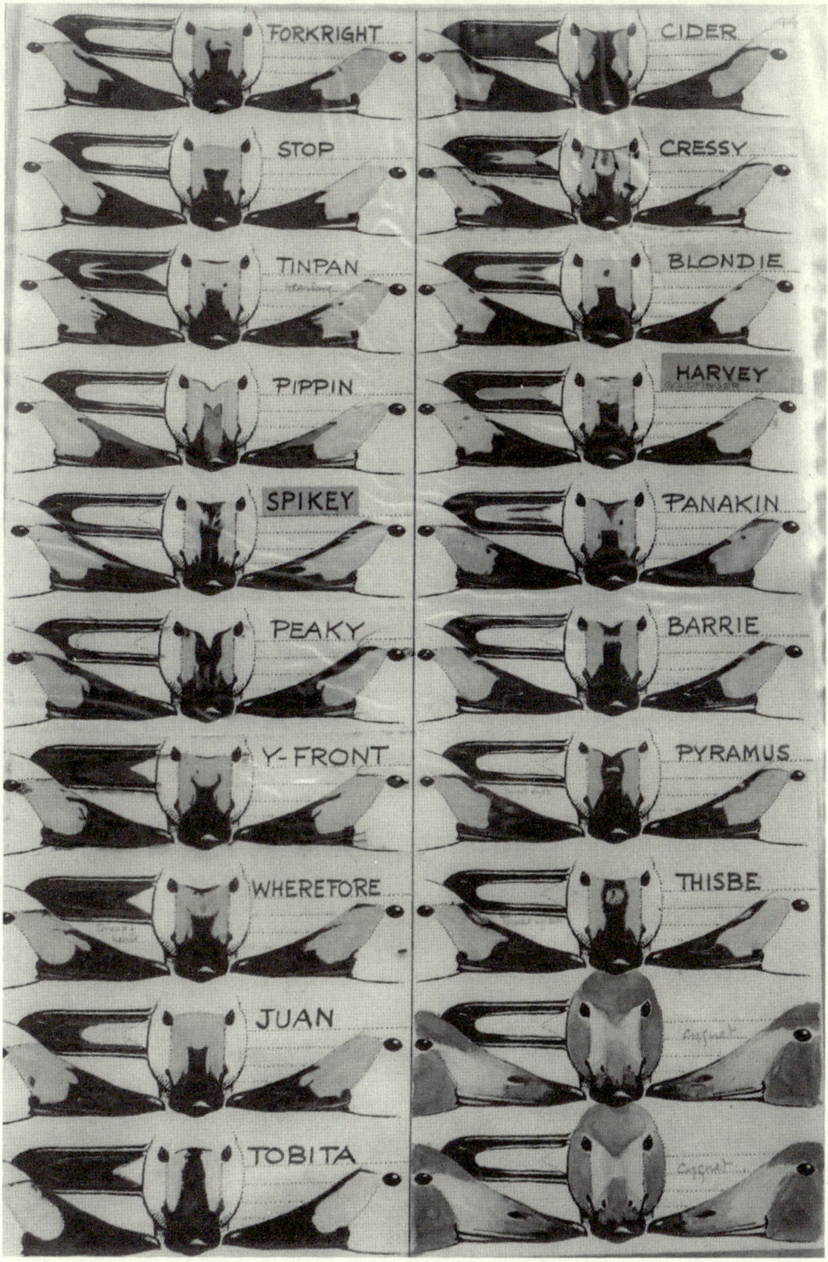

Figure 5.2 Peter Scott, Wildfowl Trust Swans (1960s).

made common cause over the creation of refuges or sanctuaries for wildfowl and accepted the legitimate harvesting of the birds where a natural surplus existed. They may have ultimately organized their respective passions for wildfowl through different technologies and acts of love (the gun rather than binoculars; observation rather than killing), but they inhabited the same 'animal landscapes' and shared the close attention to these species of birds.

Bird protection in the mid–late twentieth century did owe something to the more honourable traditions of sporting naturalism, not least from within its own ranks, with sportsmen-naturalists like Geoffrey Dent and sporting converts like Peter Scott integral to the shaping of bird conservation in the first half of the twentieth century. If a growing ethical divide began to break apart these historical ties, driven by the changing social composition of birdwatching, it could not undo the constitutive relationship between bird protection and wildfowling for large parts of the twentieth century and the productive antagonisms between the two forms of open-air recreation.

6

'Princes of the Air': Falconry and the Lure of Birds of Prey

No man made machine can, or ever will, synthesize that powerful coordination of eye, muscle and pinion as [the falcon] stoops to his kill... All in all falconry is the perfect hobby.

Aldo Leopold, *Round River* (1966, 6–7)

In its editorial from December 1974, *The Falconer* – the journal of the British Falconer's Club (BFC) – marked the twentieth anniversary of the passing of the Protection of Birds Act (1954). The Act had established a licencing system in which falconers could apply to the Home Office to take otherwise-protected wild birds of prey for use in their sport. In praising the leadership of the BFC for working with bird protectionists to put falconry on a legal footing, *The Falconer* was in no doubt about the significance of the new law: it represented the beginning of modern falconry in Britain.[1]

In freighting the Protection of Birds Act with so much significance for the sport of falconry, *The Falconer* was, in many ways, making an extraordinary assertion. The BFC itself had been formed in 1927 and had grown out of an earlier organization, the Old Hawking Club (OHC), that had operated from 1864 until 1926. The OHC had been instrumental in reviving falconry in Britain after an effective absence of nearly 200 years. The combination of Cromwell's Commonwealth's hostility to it, the invention of more efficient firearms for game shooting, and the enclosure of land had led to the effective cessation of the pastime.[2] Falconry in the twentieth century was a product of the late-nineteenth-century revival of the sport. The pastime could, however, legitimately claim long roots reaching back into the society of Medieval Britain and Europe, where, as argued by Mellor and by Campbell and Lack, it occupied a significant place within art and literature as well as within courtly and country life.[3] Venerable traditions of falconry also flourished in the early modern and modern period in the Middle East and the Indian subcontinent, with their origins dating back to antiquity. The revival in Britain in the late nineteenth and early twentieth centuries was also sustained by a similar upsurge of interest in the sport in Germany and by its growth as a popular pastime in the United States, particularly on the Eastern Seaboard.[4]

In insisting on the significance of the Protection of Birds Act to the formation of modern falconry in Britain, *The Falconer* was making the serious point that the mid-1950s marked a new chapter in the longer history of falconry in Britain – one defined by falconry's relationship with organized bird protection. This relationship had its roots in the inter-war period and had seen falconry present itself as a friend and ally of the growing campaign to protect wild birds. The sport nevertheless remained a minority pursuit in the first half of the twentieth century. In including a chapter on falconry in his celebration of inter-war 'sporting adventure', James Wentworth Day had noted that, for all its charm, falconry was the sport of 'a handful, an antique play, a diversion for men with Oxford minds'.[5] The BFC, he recorded, had no more than fifty members, though this was, in truth, a generous rounding up: as the first edition of *The Falconer* from May 1937 made clear, the club had only thirty-six full and six honorary members.[6]

As Wentworth Day's observations on falconry intimated, men from elite social backgrounds constituted this minority pursuit, with the BFC dominated by those from the landed and professional middle classes together with a smattering of military men.[7] This small 'select brotherhood' had grown to 150 members by 1959, of which about thirty owned birds. There was, however, a significant surge of interest and participation in falconry and the related ownership of birds of prey during the 1970s and 1980s. The numbers of those keeping birds of prey rose from 1,000 in 1970 to around 10,000 by the end of the latter decade. While many of the new owners would not have practised falconry in the traditionally recognized sense of flying wild, trained birds at live prey, they nonetheless represented a huge growth of interest in keeping captive birds of prey.[8]

In this chapter I explore the lure of falcons, hawks, and other birds of prey both to those attracted to the sport of falconry and to those non-falconers riveted by raptorial birds. Central to this is an attention to the aesthetic, emotional, technological, and visceral attachments made with wild birds of prey within falconry. As with the account of wildfowling developed in the previous chapter, at the heart of this is a reflection on the shaping of human conduct within the sport and its ordering of human–avian relations.

Like wildfowlers, falconers sought out intense encounters with wild birds. More so than with other field sports, however, falconry involved rigorous routines of training, care, and love for the birds. These forms of love, tenderness, and care made falconry, as Mark Cocker has noted, one of the most intimate kinds of relationship between people and birds.[9] Acts of love and care, however, were combined with the visceral pleasures of hunting and killing with them. Through these melded acts of loving and killing, falconry allowed an enlargement of the human experience, with the falconer inhabiting the world through the actions of his falcon familiar. In the process, the falconer played out multiple identifications and projections in their intense emotional bonds with the hawk.[10]

While the emotional dynamics of falconry were central to the pleasures of the sport, falconers were not unique in exhibiting intense feelings for birds of prey. They shared a strong attachment to this family of birds with the new birdwatchers. Grasping the links between these different bird-centred cultures of nature is central to my argument in this chapter. In drawing out the common attachment to birds of prey felt by falconers and conservation-minded birdwatchers, the chapter insists – as was the case with wildfowling – on the value of understanding the ways in which falconry and conservation were in a dialogue through the twentieth century. This was played out not just in relation to legislation but also across a broader cultural terrain. This imaginative dialogue variously revealed the shared repertoires through which each pastime understood and gave value to these species of wild birds. It also, however, disclosed the ethical differences that progressively segregated falconry from conservation-led birdwatching. Loving birds of prey became, by the 1970s, as much about the differences as about the similarities between the supporters of falconry and bird protection.

In the first section of the chapter I explore the rituals, routines, and practices of falconry through the first half or so of the twentieth century. The hobby was promoted and sustained not just by socially elite clubs like the BFC, but also by publishers and the wider world of field sports journalism. Through these outlets, falconry generated a significant literature both on its practicalities and more literary reflections that gave the sport a cultural profile way beyond what its actual number of practitioners might have warranted.

I explore the writings of leading falconers, most of them associated with the BFC (people such as Gilbert Blaine and Michael Woodford), together with first-hand accounts of the sport found within the sporting press. From this I move on to explore two influential literary accounts of the practice of falconry and the falconer's relationship with birds of prey. They are T.H. White's *The Goshawk*, written in the 1930s but not published until 1951, and Barry Hines's *A Kestrel for a Knave* (1968). While White's book is an account of the failed training of a hawk written by a public school and Oxbridge-educated man who was sociologically typical of the membership of the BFC, Hines's book is unusual in exploring falconry from a working class perspective. Both books are striking, however, for what they reveal about the emotional relationship between the falconer and his bird, offering insights into the enlarging of human experiences that comes from participating in the more-than-human world of falconry.

In the chapter's third section, I consider the formal relations between organized falconry and organized bird conservation from the late 1930s through to the 1970s. For much of this period, organizations such as the BFC were keen to align themselves with bodies like the RSPB and their initiatives to strengthen the protection afforded to birds of prey. These supportive relationships were strengthened in the 1960s in the wake of the 'toxins crisis' and the catastrophic decline of birds of prey within Britain, Europe, and North America. Bird conservationists and falconers acted in tandem to press governments to limit the use of the new organochlorine chemicals in agriculture and to strengthen the legal protection of all birds of prey.

This alliance began to break apart in the early 1970s, however. Fuelled by new criticism of falconry voiced at an international level within organizations such as the International Council for Bird Preservation (ICBP) and within Britain by the RSPB, falconers and falconry were subjected to a new opprobrium. This focused on the threats posed to the declining populations of birds of prey from falconers taking wild birds for their sport, and it was fuelled by a broader ethical critique of the way falconry, in keeping captive a wild bird, limited its natural modes of life. From this point on, falconry was drawn into the rogue's gallery of 'field sports' or 'blood sports', later targeted by animal welfare organizations such as the League Against Cruel Sports as much as by the RSPB.

These increasingly conflicted relations between falconers and bird conservationists were cut across by a shared way of talking about and valuing birds of prey from within both pastimes. While organizations such as the RSPB began to see falconry as part of the wider problem of raptor persecution, the ways in which leading conservationists and nature writers described and valued raptors drew on a cultural repertoire that was shared with falconry. In the chapter's final section I draw out the ways in which birdwatcher–conservationists and books aimed at them mobilized ideas about birds of prey and their appeal that overlapped with the lure of birds of prey within falconry, even as the two pastimes became more and more segregated.

THE ARTS OF FALCONRY

Driving back with his father from Winchester public school, where he was a pupil, in September 1907, Charles Portal – later to become Chief of the Air Staff and Baron Portal of Hungerford – chanced upon a group of falconers hunting partridges in an arable field. Already fascinated by depictions of falconry within books on chivalry, the scene riveted the young Portal. Alongside a number of falcons sitting on a cadge or carrying frame, he was mesmerized by the sight of Captain Gilbert Blaine bearing a falcon on his arm, the bird feeding on a partridge it had just caught. As Portal recalled more than fifty years later, 'I can still feel … the thrill of wonder and delight which passed through me that afternoon.' This was the beginning of a friendship with Gilbert (Guy) Blaine that involved, as Portal stated, 'many days of glorious sport at partridges and larks which Blaine's kindness allowed me to share with him'.[11]

At the time of the encounter with Portal and his father, Gilbert Blaine was a leading light in the Old Hawking Club. His peers saw him as one of the most gifted falconers practising the sport. In the 1920s and 1930s he held the position of vice-president of the Old Hawking Club's successor organization, the BFC, acting as an important champion and publicist for the sport. His practical knowledge of falconry, acquired over more than forty years of active hawking, was distilled into his book *Falconry*, published in 1936. The book was an introduction to and justification of the art of falconry. In its opening pages, Blaine set out to explain the appeal of the hobby. Its charm lay, he suggested, 'in its hazards, in the beauty and character of the birds themselves, in the marvellous qualities with which Nature has endowed them, and in the art of control over one of the wildest and proudest of living creatures'.[12] Managing these hazards and controlling wild birds of prey was, by necessity for Blaine, the preserve of a minority that had the necessary temperament, time, and resources to devote to the sport. Falconry, he asserted, 'can never become a popular sport', and the true practitioner could not leave the care of their hawks with the 'gardener or the chauffer'. It was also a hobby not to be recommended 'to our friends across the Atlantic', who lacked the patience to stick with the disciplined training of hawks.[13]

With these warnings set out in the introduction, and his prejudices aired, Blaine went on to describe the basic features of acquiring, training, and flying birds of prey. Using the general collective term of 'hawks' to refer to both the long-winged falcons and the short-winged hawks used by falconers, Blaine evaluated the respective virtues of each of the birds of prey typically used in the sport.

Within this pantheon of birds, one stood out for Blaine: the peregrine falcon. 'From time immemorial,' he noted, 'this has been the bird of falconry, the most universal and most useful. It seems to have been created for the service of man. Its courage, docility, and the brilliance of its performance cannot be rivalled.'[14] Peregrines also had the advantage of being easily available, with the coasts of Britain providing a ready supply of them. Blaine explained to his readers how to judge when the young peregrine was ready to be taken from the nest and how to begin the process of taming and training the young birds. Blaine called them 'eyasses' – an old falconry term used for wild caught juvenile birds. It formed part of the antique vocabulary and nomenclature used by Blaine and bequeathed to later falconers to describe the sport. Along with eyasses, there were passage hawks, caught on their first migration, and haggards, older adult birds.

This nomenclature of the different categories of birds of prey was matched by the archaic terminology used for the range of furniture or devices used to keep and control captive birds. These included the mews, where the hawk was housed; the weathering shed and weathering ground, an outside or semi-covered arena; and the baths, blocks, and perches, including the cadge for carrying hawks in the field. In addition, there were all the straps, lines, and devices applied to the bird's body to enable it to be controlled and flown. Central to these were jesses, strips of leather attached to the bird's legs; the bells sewn to jesses to enable the birds to be followed; hoods to cover their heads; and the lure and creance. The latter was a long length of cord on which the bird was flown as its training progressed.

Blaine's account of the training – or what he called the 'education' – of young birds made use of these devices and restraints. Taking around three weeks to accomplish, the process of turning a wild bird into a tamed and trained one made use of sleep deprivation and the control of food to discipline the hawk. Hoods were also central to the training and flying of the birds. As Blaine noted, 'no man can claim to be master of hawks until he is master of the art of hooding'.[15] Hoods were important in calming down the birds and giving the falconer control over them. The hoods also added to the spectacle of the sport and were often decorated with plumes. Training itself was a lengthy process of first getting the bird to sit and feed on the falconer's glove, then learning to fly to lures, and eventually to fly free and return to the falconer.

Once trained, the bird could be flown in pursuit of wild prey. Blaine described the two dominant forms of 'hawking' practised in the inter-war years. The first was 'rook hawking', carried out in open downland and farmland in southern Britain against rooks. The second was 'game hawking', which involved working with dogs and hunting down partridges (in England) or grouse (typically in Scotland).[16] Game hawking was, for Blaine, the most difficult form of hunting, requiring the birds to be unhooded and released and to wait high up in the sky before a dog pointed to and then flushed the game birds. The falcons could often be in the sky for more than half an hour and cover large distances. Not infrequently they would disappear and be lost. It was one of the 'hazards' of the hobby noted by Blaine.[17] For all its difficulty, however, it was clear that Blaine felt game hawking was a particularly fine sport. His book proudly recorded the kills made his by prized bird 'Lundy III' while game hawking. Lundy III, a male (or tiercel) eyass, bagged or scored 510 head of birds in seven years, the majority of kills being partridges and the minor fraction grouse.[18]

The naming and individualizing of the hawks, together with the recording of kills as a mark of sporting success, were integral elements of the falconry described by Blaine. The naming not only signalled the proprietorial relationship to the captive bird, but also suggested the emotional connection between the falconer and his bird, seen as a unique being with its own avian individuality and personality.

The mixture of control and the untamed power of the birds were also important to the pleasures of the sport. This was linked, in Blaine's account, to the vicarious enjoyment, through the birds, of acts of killing. As Blaine summarized it in his concluding thoughts on the sport: 'There is a strange fascination in being able to control the courses of a bird through the air; of being in touch with your falcon when she's a mere speck overhead; to witness her stoop, the swiftest spontaneous living action; to hear its rush and the heavy thud as she strikes her quarry. It is all over in a few seconds, but an intense thrill has gripped you… As she steps onto your glove from the dead quarry … you can scarcely realize the energy and power that had possessed her but a moment ago.'[19]

Blaine's book was an important touchstone for other inter-war and post-war writing on falconry, and its influence was felt in Michael Woodford's 1960 *Manual for Falconry.* As one of Blaine's successors within the upper echelons of the BFC, Woodford sought to offer an updated handbook for those interested in falconry in a more affluent age; the age of '5-day weeks, shorter working days and longer holidays'.[20] Like Blaine, Woodford offered a guide to the arcane terminology and language of falconry for an uninitiated outsider, providing a ten-page glossary of terms at the end of the book as well as a detailed chapter on the 'appliances and furniture' of falconry.

Also like Blaine, he combined a delineation of the attributes of the falconer with an account of the qualities of a good hawk. Part of the ambition in the former regard was to break with the ruthless and self-conscious elitism of Blaine's 'plea for falconry'.[21] The chapter on game hawking, culled from the notes of Major S.E. Allen, however, reinforced a message of exclusivity, noting that the falconer needed access to more than 100 acres of land and had to be able to spend six weeks during the early autumn pursuing his sport. Woodford conceded that few were likely to be in this position and would therefore be unable to enjoy the opportunities that Allen and Blaine had enjoyed in the inter-war years in owning and having access to large tracts of land and 'plenty of time'.[22]

It was not just access to land and spare time that precluded many from taking up falconry, either. For Woodford, the pastime demanded particular subjective attributes. He defined these as 'the manly virtues of restraint, perseverance, tolerance and skill'. Blaine had highlighted similar personal qualities, emphasizing 'patience, diligence [and] an even temper'.[23] Taken together with the requirement to have time and access to land, both Woodford's and Blaine's models for the human attributes of the falconer seemed to confirm the view of observers from the United States that British falconry was 'very much a gentlemen's sport'.[24]

Woodford's post-war manual repeated another dimension of Blaine's inter-war revivalist text. This concerned a central part of the pull of falconry for its practitioners: the aesthetics of the birds on the fist and in the air. If falconry was best defined for Woodford as 'the pursuit of natural quarry in natural surroundings', at the core of this pursuit was the aerial encounter with prey. Taking the peregrine as the epitome of falconers' birds, Woodford suggested that it was 'undoubtedly the speed and style of her flight which has thrilled falconers through the ages'.[25] In this sense, as other falconers regularly insisted, it was the quality of the flight that mattered more than the number of quarry species killed. During the 1950s, 1960s, and early 1970s, *Country Life* regularly devoted space to first-hand accounts of the pleasures of falconry from those practising the sport. All of these accounts celebrated the speed at which peregrines pursued their prey, with the sight of a 'stooping' peregrine arousing intense excitement. As falconer Ian McNeill put it, 'There is not, and never has been, anything quite so spectacular as the stooping falcon, our native peregrine.'[26] The bird's hunting technique was at its most enjoyable when peregrines were hunting with the falconer. Waiting for the captive bird to stoop generated intense anticipation of the drama of the chase and the kill. As Geoffrey Armitage suggested, 'a moment of real excitement comes when a falcon is waiting on high above a pointing dog and the dog is given the order to flush the game concealed in front of it'.[27]

It was not just the hunt or pursuit that was celebrated by falconers: there was also the moment of the kill. Falconers' accounts were candid about the visceral violence of this moment. In a 1963 article on 'Grouse hunting in Caithness' from *Country Life,* John Stevenson was matter-of-fact about the 'killing power' of peregrines. Describing a grouse hunt, he spelled out the moment of the kill: 'The spaniel goes in, the grouse are flushed, and the falcon, folding her wings, stoops at her chosen victim. The force of the peregrine stoop has to be seen to be believed. On one occasion a grouse was bisected in mid-air, the head and body falling some yards from each other in the heather.'[28]

The atavistic appeal of the hunt and the kill sat alongside the pull of the wild landscapes in which the birds were hunted. Captain Charles Knight – who hosted an annual hawking holiday every August at Avebury in Wiltshire (and later on Bodmin Moor) in the early 1930s, to which leading figures within the BFC were invited – captured this intermingling of the wild power of the hawks and the landscapes in which they were found and hunted. As he put it, 'there is something so purposeful, so dramatic about their shape, movements and keen expressions, something which the wild unfrequented surroundings of their natural haunts lend an added charm'.[29]

This love for wild, 'untamed' British landscapes made the moorlands of the Scottish Highlands particularly appealing for many falconers, with parts of Caithness much favoured for hawk gaming. The open landscape in Wiltshire, surrounding Avebury, was also an important locality, and it formed the 'spiritual home' of British falconry in the inter-war years. In their association with this area, British falconers shared with rural writers between the wars a fascination with the town of Avebury and its surrounding countryside as the symbol of an archaic England.[30]

James Wentworth Day picked up on this connection between falconry and 'ancient British' landscapes in his celebration of the hobby in *Sporting Adventure*, describing how falconry was played out 'on the hump-backed vallum of a British earth-work'. However, the historicism of falconry in Wentworth Day's account ranged from the Megalithic to the Middle Ages, with hawking also depicted by him as a 'pretty play of medievalism'.[31] Many falconers shared Wentworth Day's association between falconry and a 'romantic' medieval world, and they made much of its antiquity and anti-modern character.

The enthusiasm for medievalism among British falconers drew them into an association with another flourishing falconry culture in Europe with its own links to a cult of the Middle Ages: the German Falconry Association (Deutscher Falkenorden), formed in Liepzig in 1923. Renz Waller, the association's president and driving force, had succeeded in obtaining a hunting licence for falconry in 1935. He became close to Hermann Goering within the Nazi high command, cultivating Goering's interest in falconry and gaining his support for the establishment of a State falconry mews. In the late 1930s Goering sponsored an expedition to Greenland to acquire the white morph or colour form of the gyr falcon. Six of these striking birds were duly acquired. Goering planned to introduce them into the German Alps, where he believed they had once bred. The project was never undertaken, but Waller did paint a portrait, titled 'Polar falcon', of one of the birds for Goering.[32]

Falconry formed part of Nazi medievalism in the 1930s. This medievalism drew upon an already-established cult of the Middle Ages within German regional culture, especially in Bavaria, which had been mobilized in the creation of the idea of a German nation-state. Castles, ruins, and cathedrals, alongside medieval *Heimattage* ('Heimat days') involving locals dressed in medieval costume, were key elements in this pre-Nazi project of regional identity and nation building.[33] After 1933 the Nazi party appropriated elements of this medievalism, linking medieval Germanic culture to ideas of contemporary German racial superiority. Castles, including the one at Nuremberg, became the stage, as Link and Hornburg have shown, for parades and Volkish rituals.[34]

These medievalist and Volkish elements of Nazi cultural politics were evident at the International Hunting Exhibition organized by Goering in Berlin in 1937. The event included a display of falcons flying to the lure, and an exhibition of falcons and falconry furniture. The German Falconry Association's exhibit took centre stage, but there was also a British section at the event organized by the BFC, assisted by Phyllis Barclay-Smith of the International Council for Bird Preservation (ICBP).[35] The vice-president of the club, Gilbert Blaine, and its honorary secretary, Jack Magrovordato, attended as representatives of the BFC, and they brought home with them a bronze falcon statuette as the Prize of Honour for their exhibition.[36] The falcon statuette made regular appearances at the BFC's club dinners in the late 1930s.[37]

After World War II, falconers in Britain continued to draw on the revivalist tendencies and proto-medievalism of the sport's inter-war promoters, while not dwelling on its fascist associations. They regularly presented their sport as an antidote to modern life.[38] For falconer W. Kennett Richmond, the post-war interest in falconry was

Figure 6.1 Falconry display, Windsor Great Park (1960s).

certainly a reaction to technological change: 'In a push-button age, bemused with supersonic bangs', the simpler pleasures of hawking had 'an intense appeal'.[39]

Peregrine falcons in particular embodied a 'natural beauty' and 'wild nature', placing them at the margins of urban modernity. The aristocratic image of the bird and their status as a top predator were also important to falconers. For Kennett Richmond there was much joy to be had in possessing 'one of the princes of the air'. It conferred upon the falconer the values of nobility projected onto the bird. Kennett Richmond's language invoked the much-cited linking of birds of prey with social rank found within the medieval texts, where an eagle was suitable for an emperor, a peregrine for a prince, a goshawk for a yeoman, and a kestrel for a knave. The text, attributed to the Boke of St Albans (1486), was conjured early on in T.H. White's book *The Goshawk* (1951).

White was part of the elite world of falconry in Britain in the 1930s and 1940s, sharing many of the anti-modern impulses of the sport's other practitioners. He was also an enthusiastic medievalist, famously reworking Malory's 1485 story *Le Morte D'Arthur* across a number of books, including *The Sword in the Stone* (1938).[40] Both *The Goshawk* and White's practising of falconry reveal the mix of modern and anti-modern elements within British falconry in the twentieth century. They also do something else: they offer insights into the subjective appeal of falconry and, in particular, the emotional dynamics between a person and a wild bird within the sport. In the next section I reflect on White's book and his life, and I place his account of falconry next to another influential falcon tale of the post-war years: *A Kestrel for a Knave* by Barry Hines.

WHITE'S TALE AND THE STORY OF KES

T.H. White (1906–1964) – schoolteacher, author, sportsmen, pilot, and driver of fast cars – gained an international reputation for his retelling of the Arthurian legend in *The Sword in the Stone* (1938) and the associated series *The Once and Future King* (1958). Born in India to parents from the English administrative elite, White was educated at Cheltenham College and Queens' College, Cambridge. He taught English at Stowe public school between 1932 and 1936, and during his time there he began to write. Drawing on the hunting, fishing, shooting, and flying diaries that he assiduously kept while at Stowe, White gained his first success as a writer with *England Have My Bones* (1936). The book was a celebration of mainly rural pursuits, including wildfowling, though it included learning how to fly a plane and driving fast cars.[41] The book evidenced White's compulsion to learn new skills. Asked while giving a talk in later life to say something about himself, he saw this desire to learn things as a way of 'compensating for my sense of inferiority, my sense of danger, my sense of disaster'. In the same talk he had quoted Merlyn from *The Sword in the Stone*, where the wizard had proposed that 'the best thing for being sad is to learn something'.

As White's biographer Sylvia Townsend Warner noted, so much obsessive learning from White 'presupposed a good deal of sadness'.[42] Behind Townsend Warner's observation lay White's troubled childhood and his lifelong struggles with his homosexuality and his own sadistic fantasies. White turned to psychoanalysis to help him with his inner demons. Therapy sessions often left him elated: 'I am so happy that I hop about like a wagtail in the streets,' he confessed in 1935. 'P.A. [psychoanalysis] is the greatest thing in the world.' Buoyed by the benefits of therapy, he toyed with publishing his own 'sexual autobiography', 'for the benefit of other poor devils', though he never produced it. He did, though, keep detailed notes on his mental state later in life.[43]

One activity that fitted White's obsession with learning new skills and which spoke to his inner turmoil was the training of hawks. Spurred on by a passage in a book on falconry that he had come across, White gave up his teaching job and rented a labourer's cottage on the edge of the Stowe estate, writing to Renz Waller in Germany to ask for a young goshawk to train. White's account of his life in the cottage and the traumatic story of the ultimately failed attempt to train the bird he called 'Gos' was written up from his daily records of the 'manning' or training of the bird. White put the manuscript aside and it was not until 1949 that his publisher came across it by chance and persuaded him to publish it. White was hesitant but consented, and the book, *The Goshawk*, appeared in 1951, largely to critical acclaim. It did have its critics, however. A reviewer in *The Falconer* warned that it was effectively a catalogue of 'things one should not do' in training a hawk.[44]

The Goshawk followed the diary-style format from which it was derived, mixing the events of days or groups of days with wider reflections on falconry, the modern world, and White's own emotional state. The book is a painful read, as White battles to train his hawk, withstanding weeks of sleeplessness, physical discomfort, and small successes mixed with many failures in the training of the bird. White followed the guidance of Gilbert Blaine's book on falconry together with a 1619 *Treatise on Hawks and Hawking*.[45] The 1619 tome proved not to be a good choice, and it contributed to his failure in training Gos, with White conceding that it provided 'a very out of date idea of the way to man a hawk'.

For all the shortcomings of his attempt to train Gos, White's account is rich in its reflections on his relationship with his hawk. The book documents the great patience, love, and care for the bird that are essential to the 'manning' process. This emerges most clearly in White's account of the early stages of training Gos. A recurring problem for White was that the bird would regularly struggle while being held on his gloved fist, moving from a state of apparent contentment to one, as he put it, of 'rage and terror'. This resulted in the bird 'bating'. As White described it: 'The leashed hawk leaps from the fist in a wild bid for freedom, and hangs upside down by his jesses ... like a chicken being decapitated, revolving, struggling, in danger of damaging his primaries.'[46]

The falconer had to respond with quiet care and love to the bird in its distress, lifting it back onto the fist with 'gentleness and patience, only to have it bate again, once, twice, 20, 50 times, all night'.[47] In this battle with Gos and the moves towards eventually being able to fly the bird in the field, White reflected on the shifting relations between the falconer and the bird during the power play of training. As the days of what White called 'attack and counter attack' unfolded, he wondered whether he was the 'master' or whether it was Gos that was in charge. He thought he knew the answer: '"Gos" regarded me with tolerant contempt. He had no doubts about who was the slave, the ridiculous and subservient one who stood and waited.' White pityingly compared himself to a butler, waiting on his master and adopting the butler's 'distant gaze'.[48] He conceded that he had been subjugated to the bird's 'brutality' – this bird in whose 'talons there was death [and which] could slay a rabbit in its grip by merely crushing its skull'.[49]

The powerful bird could switch from being a fierce predator to being a comic child: 'funny and silly little Gos'.[50] Drying itself after its first bath, the fluffed up hawk then turned into a music hall character, 'the princeling ... the infant Tarquin had suddenly become a charlady at Margate', hitching up her skirts. Sometimes these shifts in the positions occupied by Gos were abrupt. The bird could go from being 'a homicidal maniac' one moment to placidly enjoying being stroked. In this switch of mood, White noted, 'we were in love again'.[51]

White would often see Gos, as Helen Macdonald has argued, as an unruly public schoolboy in need of education – one that had to be 'broken' from his 'ungovernable state' into one that had been successfully trained.[52] White was also prompted to see himself as the bird's mother as much as its teacher. The effort and worry occasioned by the training of the hawk over many weeks was like 'a mother nourishing a child inside her'. Like the relationship between mother and child, the training formed a permanent bond with the bird. Man and bird became, for White, 'linked by a mind's cord'.[53] White's sense of himself as occupying this maternal role was reinforced in *The Goshawk* by the domestic routines that filled his life during the training of Gos. He painted the house, made bookshelves and a table, scrubbed floors, cleaned grates, and shot rabbits and pigeons for Gos.[54] But for all his love and devotion towards the bird, White's attempt at training failed. While flying Gos on a creance, the bird broke free and flew away. White searched frantically for it, briefly seeing it again before it disappeared for good. A dark image overcame him. He pictured Gos hanging dead, its jesses caught in the branches of a tree, its feathers matted and its body putrefying.

The heartbroken White sought consolation in his garden and surrounded himself with other wild animals that needed to be cared for. Extending his mothering role, he confessed that these were all animals that had to be fed, like the young badgers that he gave warm milk and sugar to out of an old Champagne bottle.[55] If White filled his life with significant animal relationships, he continued to see hawks through shifting projections. He described watching a young peregrine on the ground through binoculars, 'flapping his wings like a small pre-school boy doing Sandow exercises'.[56] In his diary for 1938 he attached under cellophane a tiny lark's claw with the inscription 'Red's first lark', as if the bird was a child and he the parent keeping a record of its achievements.[57]

White's paternal/maternal feelings for birds were perhaps only trumped by his feelings for his dogs. When his red setter 'Brownie' died, he wrote in his diary that it was the end of '12 perfect years of love'. The dog had been, as he put it, 'the central fact of my life'. A year after its death, he was still mourning: 'My life is nothing, my darling, without you,' he recorded.[58]

The emotional relationships forged by White with his companion animals, including his hawks, and the solace they offered him found echoes in the story of 'Kes' and its owner Billy Caspar in Barry Hines's (1939–2016) *A Kestrel for a Knave*. The story of Kes, like that of Gos, is one with a tragic ending, with the young kestrel Kes dead in a dustbin by the end of the story. But at the book's heart is the relationship between a troubled, misunderstood schoolboy, Billy Caspar – a 'lonely misfit' as Hines later described him – and his trained kestrel, and the solace and joy Billy finds in his training and flying of the bird.[59]

The book is set in a Northern working class mining town based on Hoyland Common near Barnsley, where Hines grew up.[60] Hoyland

Common was a semi-rural community, with the mine and the town surrounded by woods and fields. The area was one in which, as Hines powerfully noted, 'the miners walked to work across meadows, with skylarks singing overhead, before crowding into the cage at the pit top and plunging into the darkness'. This juxtaposition of industry and countryside informs the landscapes of *A Kestrel for a Knave.* Billy moves between the grey streets where he lives with his mother and brother to the open country and woodlands nearby.[61] Billy straddles these two worlds, moving with particular ease and attentiveness through the non-human world around him in the fields and woodlands on the outskirts of town.

The setting of *A Kestrel for a Knave* is very different from White's rural idyll in *The Goshawk,* and the storytelling in each book is also radically different. While *The Goshawk* follows the relentless style of a diary in unfolding its tale, *A Kestrel for a Knave* is set on one day, with flashbacks to earlier episodes. Hines also used Northern dialect in the book, seeking to capture the rhythms of working class speech and intonation. Billy's family life with his mother and brother is shown as disorganized and violent, with Billy subject in particular to bullying from his brother. In a household without a father present and with a neglectful mother, the book, as Ian Haywood and Dave Russell have both argued, presents an unromantic depiction of working class life and is distanced from those fictions that have foregrounded working class solidarity and the culture of traditional labour. The injuries of class, however, are central to *A Kestrel for a Knave,* with its most withering attacks focused on of the damage done to working class children by selective education and secondary modern schooling. The cruelties and brutalities of this education are shown in Billy's experiences of school, and the novel clearly sets out to show how the system of education is designed to fail working class pupils, preparing them only for low-skilled jobs. *A Kestrel for a Knave* is, in this regard, a literary precursor to Paul Willis's *Learning to Labour.*

Billy's interest in falconry and his relationship with Kes allow Hines to reveal Billy's potential: his ability to master the complex nomenclature of falconry and to excel in ways that he had not been able to at school. In one of the book's central scenes, Billy, supported by the sympathetic schoolteacher Mr Farthing, takes charge of the class and explains the terminology of falconry to a rapt classroom of his peers. Drawing on a handbook of falconry that he has

Figure 6.2 David Bradley as Billy Casper in the film *Kes* (1969).

stolen from a bookshop – and which is depicted in the film of the book as Michael Woodford's *A Manual of Falconry* – Billy regales the class with his knowledge. As he says: 'You train 'em through their stomachs… You call this manning. That means training.' He also details key parts of hawking's furniture, including jesses and the creance.[62]

Impressed by the boy's expertise, Mr Farthing meets Billy after school and watches him demonstrate his falconry skills. Billy's mastery of the art and his connection with the bird is captured in the scene: 'Working the lure like a top matador his cape. Encouraging the hawk, making her stoop faster and harder, making Mr. Farthing hold his breath at each stoop and near miss.'[63] Billy's ease and connection with the bird is evident, and he finds a new kind of expression of himself as he works with Kes, as the bird flies to the lure. His care and attention is also evident in the way he shoots sparrows to feed to Kes, gently holding the food on his fist while Kes eviscerates it and gobbles it down.

White might have seen it as mothering, and it is certainly a relation of care and love for Billy too. As it was for White, Billy's love for the bird is part of a wider connection with and feeling for wild animals. He tells Mr Farthing that he has kept 'stacks of animals – young foxes, magpies, jackdaws, jays'. For Billy, though, Kes is definitely not a pet. As he says: 'They can keep their rabbits an' their cats an' their talkin' budgies, they're rubbish compared wi' her.' It is Kes's residual wildness and its fierce nature that thrills Billy. As he forcefully tells Mr Farthing: 'It's not a pet, sir, hawks are not pets… Is it heck tame, it's trained that's all. It's fierce, an' it's wild, and' it's not bothered about anybody, not even about me right. And that's why it's great.'[64]

Tony Garnett, the producer of the film *Kes*, saw this line as Billy 'talking about his own class' and 'as a statement about working people'. But we may more profitably read it as being about Billy recognizing the separateness of Kes and its lifeworld and thrilling at being connected to its more-than-human world. While Billy identifies with the independence and wildness and violence of the kestrel's life, he does not want to dominate it and own it in any proprietorial sense. It is about a part of Billy 'becoming Kestrel'.[65]

FALCONRY AND CONSERVATION

By the time *A Kestrel for a Knave* was published, it was illegal to take any British bird of prey from the nest. Sparrowhawks had recently been added to the list of wild birds given full legal protection, though it was still possible to apply for a licence to take kestrels for the purposes of falconry. This tightening of legal protection had been occasioned by the dramatic declines in the populations of birds of prey, with the peregrine – the favourite bird of British falconers – in particularly steep decline as a breeding species.

Falconers had been among the first observers to realize that there was a problem with the breeding success of peregrines. Writing in *The Falconer* in December 1958, R.B. Treleaven, who had spent years watching the peregrines that bred on the north Cornish coast, reported that 1958 had been 'an exceptionally bad year', with the complete failure of all breeding attempts in North Cornwall.[66] He speculated that increased human traffic along the Cornish coast, together with the growth of clifftop vegetation brought about by the decimation of the rabbit population from myxamatosis, might have been responsible.

Nature Conservancy's peregrine study, published in 1963, soon revealed the real reasons for the birds' decline by establishing the link between the widespread use of new organochlorine chemicals in agriculture and the population decline of British peregrines. In the context of this evidence and the threat to birds of prey, the BFC sought common cause with the RSPB. This built on connections between the two organizations and allied groups like the Association of Birdwatchers and Wardens (ABW) going back to the 1930s.[67] In the controversy stoked in the late 1950s by racing pigeon interests that prompted Nature Conservancy's peregrine inquiry, the BFC aligned itself with the cause of the protectionists, challenging the 'emotional propaganda' of racing pigeon organizations. Supporting the need for 'official and unbiased' evidence on the status of peregrines, an editorial in *The Falconer* emphasized that falconers were as 'keen protectionists' as ornithologists.[68] The journal also strongly supported moves led by the ICBP at its 11th International Conference held in 1958 to give greater international protection to birds of prey, and to challenge the deep-rooted hostility to these birds across many countries.[69] When the RSPB and the BTO organized a conference dedicated to birds of prey in Cambridge in 1963, the BFC was an enthusiastic participant.[70]

The attention given to the dangers posed to birds of prey by toxic chemicals at the Cambridge conference, and to the need to strengthen the legal protection received by this group of birds, was taken forward by an international gathering held under the auspices of the ICBP at Caen in France. Representatives from ten Western European countries and the United States discussed the threats faced by raptorial birds and the need for coordinated international action to protect them. Jack Mavrogordato attended as the BFC's representative on the British section of the ICBP. Like the birds of prey conference in Cambridge, the proceedings at Caen concluded that greater protection of these species was needed because of the role they played in 'maintaining an equilibrium in natural wildlife populations and because they form part of our international heritage'.[71] It was a view endorsed by the BFC.

The fruitful dialogue between the BFC and conservation scientists was also maintained through the late 1960s, with both Derek Ratcliffe and Ian Prestt from Nature Conservancy giving talks to BFC members on their research on the peregrine and sparrowhawk censuses, respectively.[72] The harmonious relations between falconers and bird protectionists in Britain did not persist, however, and were disrupted by wider international debates. In 1968 ornithologists meeting in Hungary for the conference of the ICBP's European Continental Section put forward a motion seeking to ban falconry as 'an illegitimate, unjustified and anachronistic form of sport'.[73] This criticism was extended in a survey undertaken by the ICBP's European Section during 1969. The survey highlighted the concerns from French representatives about the actions of German and Austrian falconers, who were accused of taking wild falcons in parts of France and Italy.[74] The growing body of criticism of falconry within the ICBP was expressed again at the first meeting of the World Working Group on Birds of Prey, chaired by Peter Conder of the ICBP's British Section (Conder was also secretary of the RSPB). As *The Falconer* wearily reported, the conference was marked by the 'usual attacks ... against falconry and falconers'. They included a comment from the West German representative Dr W. Erz, who alleged that 'falconers did more harm to birds of prey than pesticides'.[75]

In the United Kingdom, the RSPB became more openly critical of falconers as it challenged what it saw as a growth in the persecution of birds of prey. 'Would-be falconers' were seen as part of the problem, given that the demand for falcons greatly outstripped the number of licences granted. With the bird protection laws difficult to enforce and the penalties for committing bird crime weak, falconers shifted from being allies in the cause of conservation to being culpable through their activities of threats to birds of prey. A comment piece by the RSPB's director Peter Conder in the winter 1970 edition of *Birds*, the RSPB's members' magazine, developed this new, critical line of attack. While Conder conceded that the RSPB was not against 'legitimate' falconers, he warned that a growing interest in falconry was drawing in those without proper training in the 'ancient art', who were unable to look after the birds properly. The piece was not just critical of the 'lads who shin up a tree to take a young Kestrel', but was also scathing of commercial falconry displays. Not only did these displays encourage the untutored to seek out hawks and thus put pressure on the wild populations, but they were also undignified in their display of the birds. As Conder noted, performers turned up 'at show grounds in exotic costumes', transforming these wild birds into a circus exhibit. Even more so than legitimate falconry, they destroyed the 'elemental ethos' of these birds – birds that had, he argued, 'a fierce, free quality like the wind itself'.[76] For Conder, then, falconry displays and falconry more broadly distorted the natural modes of life of the birds. In a subsequent leaflet titled 'Falconry, should I take it up?', the RSPB encouraged those who might be thinking of trying falconry to think again. 'If one is genuinely interested in this fascinating group of birds,' it argued, 'it is better to study them in their natural environment rather than keep them in captivity.'[77]

The BFC was stung by these criticisms of its sport. The club redoubled its efforts to bolster 'legitimate' falconry and continued to seek common cause with conservation. In the June 1972 issue of *The Falconer*, the club's president, Jack Mavrogordato, defended the sport from the misapprehension that it was a major threat to birds of prey. He argued that falconers had always had an interest in preserving the stocks of wild birds of prey, and he noted how the BFC had published its own leaflet that discouraged the majority of would-be falconers from trying their hand at the sport. These were 'those whose imaginations may have been fired by the sight of a Game Fair display or the film with a highly dramatic and sentimental theme of a boy and a kestrel'.[78] Another member of the BFC, C.J. Morley, took to the letters pages of the RSPB's members' magazine to defend his sport against the RSPB's criticisms. In defending legitimate falconry, Morley targeted the 'illegitimate' falconers and those who promoted an unhealthy interest in birds of prey, such as 'zoos, falconry centres, game bird fairs and television' – all those who stirred demand for birds from the 'pseudo-falconers'.[79]

This presentation of 'true falconry' as a minority, elite pastime deeply committed to conservation was driven by a clear politics. This was to maintain the alliance between organized falconry and organized birdwatching and thereby protect the sport from any change in the law.[80] The RSPB's criticism of falconry, however, hinted at a deeper ethical divide. However well organized and committed falconers might be to the 'art' of falconry, there was a deeper problem with the practice. As critics like Peter Conder rhetorically asked, in owning and keeping captive birds of prey did falconry not take away their freedom? And as another acute observer commented, did not the hobby ultimately put pressure on wild bird populations, being, in effect, as much a threat to peregrines 'as toxic chemicals, uneducated game-keeping and serious loss of habitat'?[81] This growing ethical divide was on show in the RSPB's 1973 campaign to remind the public of the legal protection afforded to all birds of prey. In its publicity the Society highlighted the problem of 'egg thieves and would-be falconers' as the main culprits in the persistence of bird crime directed at birds of prey.[82]

FIERCE MAJESTY

In the criticisms made of falconry by organizations such as the RSPB during the early 1970s there was little or no attention paid to the bloodthirsty dimensions of the pastime. Indeed, ornithologists and falconers shared their fascination with the aerial prowess of the peregrine, the speed of its hunting stoop, and the way the birds embodied a sense of wildness and power. As Peter Conder put it in his editorial piece in *Birds* that was cited above, peregrines had a 'fierce, free-quality like the wind itself'. In countryside and bird books aimed at birdwatchers there was a foregrounding of the peregrine's aerial skills and power as a predator. As the AA's *Book of the British Countryside*, published in 1973, put it: 'A peregrine can dive at speeds … up to 180 mph, striking its prey with instant death and swirling upwards again on sickle shaped wings.'[83]

This attention to the fierceness, speed, and killing power of peregrines had a long pedigree within ornithological circles. In an RSPB pamphlet from the mid-1930s, W. Salter described the peregrine as the 'handsomest and most striking of all our birds of prey … the embodiment of speed and strength'.[84] Across the Atlantic, American ornithologist Roger Tory Peterson deployed similar language in the late 1940s. In a chapter of his *Birds Over America* (1948) devoted to the peregrine, he emphasized the dramatic power of the bird's hunting 'stoop' and the way it attained 'rocket-like speed'. In another passage Tory Peterson suggested that there was a similarity between a falcon and a 'fighter-plane or dive bomber'. In invoking these parallels between the qualities of the falcons and modern aviation and warfare, Tory Peterson was typical of a more general impulse to see peregrines through the lens of the contemporary celebration of speed and the aerial prowess of modern aviation.[85] While, as Pyrs Gruffudd has shown, in the inter-war years it was those on the political right who were particularly obsessed with these qualities of aeroplanes as part of their broader infatuation with speed, power, and modern machines, by the 1940s the modernity of aerial power and aerial warfare had a wider cultural resonance.[86] Modern aviation could also be subsumed within medievalism. Pilots, as Gruffudd suggests, were seen to embody chivalric virtues and were often caste as 'latter-day knights errant'.[87] The same courtly roles could be occupied by peregrines. Tory Peterson elaborated on his account of the falcons by drawing on the link between the birds and elevated social rank. As the New York peregrines flew high and unnoticed above the heads of shoppers on 5th Avenue, Tory Peterson saw them as the 'prince of predators'.[88]

The focus on the bird's speed and killing power, and the falcon as an analogue for a modern fighter aircraft, surfaced in other books aimed at a conservation-minded public. In a book on birds of prey, Philip Brown (former RSPB secretary) and Colin Willcock (of Anglia Television's *Survival* series) emphasized contemporary aviation parallels and the killing power of peregrines. In his foreword to the book, Willcock described his own encounter with a peregrine on the English coast and recalled the moment when a mallard was struck by a peregrine. The duck had been steadily flying along, he recounted, when in an instant it 'had become a tattered bag of feathers'. 'The peregrine,' he went on, 'in its 500 feet dive had arrived like a jet-fighter attacking a bi-plane. It was magnificent.'[89]

In his chapter on the peregrine in the book, Philip Brown also emphasized the deadly power of the peregrine as an aerial hunter. This was a bird, he noted, 'built for speed ... a bundle of feathered dynamite'. Like other ornithologists and falconers, Brown was riveted by the bird's stoop and thrilled by its killing of quarry, and he noted how the speed and power of the stoop would decapitate the falcons' victims. Such a death, he apologetically added, was surely the swiftest and least painful way to be killed.[90]

An image of a peregrine falcon striking its avian prey in mid-air – the quarry crumpled and broken, with feathers flying – featured as an illustration in the RSPB's programme for its 1970 film *Winged Aristocrats*. Released to coincide with European Conservation Year 1970, the film included footage not just of British birds of prey but also of raptors from across Western Europe.[91] The film presented an ecological message about the role of birds of prey in the functioning of natural ecosystems and how the birds were adapted to the environments in which they lived. It also discussed how these birds had been subject to long-standing persecution, ending with a plea for their greater protection.

In its title and its emphasis on the charismatic qualities of birds of prey, *Winged Aristocrats* drew on the association of many raptors with elevated social rank and on the appeal of their fierce beauty, as killers with hooked beak and claw. The film was also drawn to the spectacle and mythology of falconry itself. In its opening shots it depicted a falcon in action before moving to a close-up of the bird's head. As the camera panned back, the viewer saw that the falcon was sitting on the fist of a falconer dressed in medieval clothing. The film then discussed the history of birds of prey within falconry, mythology, and heraldry.[92] In making these connections with the cultural and sporting traditions that had shaped contemporary ideas about birds of prey, the film acknowledged the links between the long-standing human associations with them and the contemporary conservation message. In doing so, it made clear the shared imaginative space that joined these different bird-centred cultures of nature.

These imaginative connections between birdwatching and the sport of falconry were played out in J.A. Baker's celebrated book *The Peregrine* (1967). The book was notable for the way it framed the narrator–watcher's pursuit of wintering peregrines. Central to this is the metaphor of the 'hawk-hunt' that drove the narrative of the book. The idea of hawk hunting is an odd phrase for a birdwatcher to use. Baker talks in the book's framing pages less of watching, seeing, or recording peregrines than of hunting them, as if he were engaged in a non-lethal version of field sports. While his narrator is not using peregrines to hunt other birds, he shares with the falconer the pleasures of the pursuit and the chase. This gives an intensity of focus to the encounters with wild birds, heightening the senses of the 'hunter'. 'Hawk-hunting sharpens vision,' the narrator suggests on the fourth page of the book: 'Pouring away behind the moving bird, the land flows out from the eye in deltas of piercing colour.'[93]

In framing the account of his pursuit of peregrines in this way, Baker defined his quest for the birds as a process in which he sought to enter into the life-world of his prey; to begin to see the landscapes that bound him to the birds in the same way the peregrine did; to become one with the bird; to become himself a hawk. In this sense, the narrative of *The Peregrine* is not so much a testing of the narrator's self, but an act of shamanic transformation – of 'becoming-Peregrine'. This is most evident in the entry for 30 November in Baker's fictional year. Coming across the body of a wood pigeon killed by a peregrine, the narrator mimics the predator's behaviour, taking on its form: 'I found myself crouching over the kill, like a mantling hawk. My eyes turned quickly about, alert for the walking heads of men. Unconsciously I was imitating the movement of the hawk, as in some primitive ritual: the hunter becomes the thing he hunts.' Elsewhere in the book, the narrator picks up a dead gull, smelling the freshly killed flesh as if to experience how the hawk would sense it.

The recording and documenting of peregrine kills forms a central theme in the narrative of *The Peregrine*. In particular, it is the emotive, sustained descriptions of killing that are central to the visceral identification with the life-world of the bird that are rendered so strongly in the book. These descriptions also bring Baker's narrative close to the fascination with and pleasure in kills that are evident within the culture of falconry. In the chapter titled 'Beginnings', the narrator prepares the reader for what is to come: 'I shall try to make plain the bloodiness of killing,' he warns. The section titled 'Peregrine' explains how the physiognomy of the bird is adapted to the 'pursuit and killing of birds in flight'. The narrator explains at some length and in great detail the peregrine's stoop: 'The stoop is a means of increasing the speed at which the hawk makes contact with the prey. The momentum of the stoop …enables him to kill birds twice as heavy as himself.'[94] The power of the stoop and its effects on the peregrine's prey are conjured most powerfully in the description of the killing of a partridge from 10 February:

> His speed increased and he dropped vertically down… He fell sheer … heart-shaped, like a heart in flames… The partridge in the snow beneath looked up at the black heart dilating down upon him… And for the partridge there

> was the sun suddenly shut out ... the blazing knives driving in, the terrible white face descending... And then the back-breaking agony beginning ... till the merciful needle of the hawk's beak notched in the straining neck and jerked the shuddering life away. And for the hawk ... there was the rip and tear of choking feathers and the hot blood dripping from the beak and the rage dying slowly to a small hard core within.[95]

The narrator confesses to the vicarious pleasure of watching this kill, the 'memory of that sabreing fall from the sky' etched in his mind, the 'guiltless hunter who kills only through his familiar'.[96] Baker's *The Peregrine* shares the bloody fascination with the peregrine's hunting life with falconers, and it links the aerial prowess of the birds to the sense of wildness and freedom that they embody. Much of the narrative of the book is absorbed by the narrator's desire to merge with the *umwelt*, the life-world, of the peregrine, to see the world as he does and to break free from human constraints. As the narrator puts it at one point, 'imprisoned by horizons, I envied the hawk the boundless prospect of the sky'. The liquid mobility of the flying peregrine is most memorably captured towards the end of the book: 'Free! You cannot know what freedom means till you have seen a peregrine loosed into the warm spring sky to roam at will through all the far provinces of light ... like a dolphin in green seas, like an otter in the startled water, he poured through deep lagoons of sky up to the high white reefs of cirrus.'[97]

CONCLUSION

Later commentators have seen in J.A. Baker's identification with the freedom of the wild peregrine and his narrator's quest for the bird a search for personal redemption. As Helen Macdonald has suggested, *The Peregrine* represents a 'soul's journey to grace, a man looking for God'. For Robert Macfarlane, the book turns the watching of birds into a 'sacred ritual'. Through the subsuming of the self into another creature, by 'becoming a bird', the narrator, for Macfarlane, seeks to escape the troubles of his human self, to push back the reality of death, and to find a 'resurrection'.[98] Baker's projection onto the wild falcons the qualities of freedom and abandonment and the narrator's identification with these attributes certainly speaks to a desire to break free from the constraints of human existence and the world of social convention and obligation. It is a tale fuelled by misanthropy, in thrall to the clarifying power of violence, killing, and blood, and driven by an imagined flight from the old self into a new one – by the salvation granted by 'becoming peregrine'.

Such a reading of *The Peregrine* and its understanding of human–avian relations represents a view of natural relations that comes close to seeing them as a compensation for the dysfunctional aspects of human relations and as solace from social marginalization and inner conflicts. It is an account of human–falcon relations that finds echoes in those two other influential mid- to late-twentieth-century tales of birds of prey and their human companions: *The Goshawk* and *A Kestrel for a Knave*. As we saw, both T.H. White and Barry Hines presented the relationship of their central characters to a bird of prey as offering comfort and solace from the inner and outer troubles of their lives. The role of a falconer allowed a close and intense relationship with a wild bird – one that offered the falconer the possibility of playing the role of master, carer, lover, teacher, mother, companion, and servant of the bird.

The bird's latent power and capacity for violence heightened White's own emotional states. He took pleasure in its killing and wished himself to become a wild and feral thing like 'Gos'. In doing so, White was – like Baker – seeking escape from the traumas of the world and his own demons. Billy Caspar's escape was more limited, but was rooted in his experience of class and the constraints placed upon his life chances by his circumstances. Like White, Caspar acquired new skills and learnt new things that enabled him to accomplish more than he had been able to at school. The companionship with 'Kes' enlarged his experience, allowing him to express himself in new ways and opening him up to elements of the lifeworld of birds.

While it is important not to treat the stories of White and Gos and Billy Caspar and Kes as representative of how all falconers lived their relationship with their birds, both books offer insights into the wider culture of falconry in twentieth-century Britain. The routines of care and love, and the shifting roles of master and servant, undoubtedly speak to how other falconers experienced the sport of falconry. As we have seen, many were, like T.H. White, men from elite social backgrounds, and they shared similar relations with hawks. As Helen Macdonald has argued, these men seem to have been drawn, like White, to the sport because it also allowed them to project onto the birds qualities they felt were missing from contemporary life or under threat from it.[99] The antiquity of the sport, its romantic medievalism, and its association with wild, non-urban and sometimes 'ancient' landscapes fed their antediluvian impulses. This was a sport for anti-modern individualists, as an editorial in *The Falconer* conceded. It did 'not set out to be a cosy little suburban gathering with chit chat over the teas cups or pint pots [and] it has no wish to steal the thunder from the local tennis or golf club'.[100] It was a sport for men of initiative, it argued, even in the 'era of the Welfare State'.

The position of birds of prey as predators at the top of food chains also encouraged an identification with them. Food chains could be read across to other, social, hierarchies, with social status accruing to the birds and those who possessed them. The old medieval texts licenced this interpretation, with birds such as peregrines and eagles used to signify high social rank and endowing the birds with nobility. As 'winged aristocrats' and 'princes of the air' these birds allowed those that flew them, or even simply watched them, to possess some of their qualities. In this regard, falconers and ornithologists not only purposefully sought out similar encounters with wild birds of prey, but were also drawn to project onto the birds similar attributes. There were other connections. Falconers may have used different technologies and devices to bring birds close to them and to control them and hunt with them, but they also used binoculars to track birds and were as attentive observers of their behaviour and plumage as ornithologists.

It was in relation to the capture, containment, and control of the birds that ethical differences divided the two bird-centred cultures of nature. Certainly by the 1970s, leading bird conservationists like Peter Conder were highlighting the discomfort they felt about the breaking of a bird's natural modes of life within falconry. Gilbert Blaine had spelt out the accomplishment and pleasure he took from controlling a wild bird in the 1930s, but for Conder and others it was this control that was at the heart of their criticism of even 'legitimate' falconry.

Some of falconry's critics were more direct even than Conder. As Helen Macdonald has revealed, one such critic wrote to T.H. White after reading *The Goshawk.* He berated White, shocking him in the process, for the way he had treated a bird White claimed to love.

As White's critic suggested: 'How can you talk of love for a bird after subjecting our wonderful predatory birds to such torture… Is there not enough cruelty in the world without adding to it for one's amusement or hobby.'[101]

For organizations such as the RSPB and the ICBP, it was additionally the link between falconry and raptor persecution that provoked their criticisms. These different dimensions of the critique of falconry fed into a more sustained investigation by the animal welfare organization the League Against Cruel Sports in the early 1990s. The resulting report – written by Peter Robinson, the former senior investigations officers of the RSPB – highlighted the link between the desire of many people to own birds of prey and the high levels of nest robberies. The report also raised concerns about how falconers and other owners exercised control over the birds they held captive. In a final observation, the report reflected on the appeal of owning a wild bird, suggesting that for many, particularly those who liked to be seen with a bird of prey on their arm in public, there were likely to be feelings of 'personal insecurity or inadequacy' lying behind their choice of hobby.[102] T.H. White might have agreed with the latter observation, but it was nonetheless a cheap shot at the appeal of falconry and of wild birds of prey in general. The insinuation of 'personal inadequacy' was also, as we have seen, a simplification of the complex emotional dynamics played out between people and hawks within the sport. More than that, it missed the shared feelings and cultural values through which both falconers and conservationists responded to the lure of birds of prey.

7

Harvesting the Sea and the Air: Wild Birds, Food Customs, and Conservation

In October 1969 the *Sunday Mirror* ran a report from its investigative team about the mass killing of migratory birds in Belgium. Under the headline 'Massacre of the songbirds', the article detailed the 'orgy of destruction' committed by Belgium bird-trappers. Using live bird decoys and nets, 'millions of homely birds ... the sort of birds that you welcome to your garden as friends' were trapped, and their 'tiny carcass[es] prepared for a gourmet's meal'. Paul Hughes, the lead reporter, confessed to being shocked as trappers talked casually of slaughtering songbirds. Noting that in Britain and 'most civilized countries' this kind of trapping was illegal, the paper detailed the 'horrifying story of the bird trappers'. Alongside pitiful images of decoy birds held in nets or suspended on a line to attract other birds, the report built up in brutal detail the accounts of killing from the mouths of the bird hunters themselves, showing how they squeezed the life out of birds by crushing their heads between forefinger and thumb.[1]

Eighteenth months earlier, *The Guardian* had run a similar story about the killing of songbirds in Italy. Under a headline that foreshadowed the *Sunday Mirror*'s, the paper documented the annual killing of 150 million birds in Italy, the overwhelming majority of which were migrating songbirds.[2] Two years later, in February 1970, *The Times* also ran a story about the reintroduction of bird netting in Italy. Prompted by this piece, reader H.W. Clarke, from Staines in Middlesex, wrote to the director of Nature Conservancy (NC) seeking the support of his department in bringing pressure to bear on the Italian government concerning what Clarke called 'the barbarous netting of migrant birds'. If nothing was done, he argued, 'many of our most attractive birds will disappear for ever'.[3]

In drawing attention to the trapping, killing, and eating of wild birds in parts of northern and southern Europe, the British newspaper reports and at least one of their readers unequivocally condemned these Continental practices, differentiating the laws and behaviour of 'civilized countries' like Britain from those on the other side of the Channel. The *Sunday Mirror*'s report in particular, but also the response of H.W. Clarke, additionally sought to stake a proprietorial claim on these songbirds. The birds needed to be protected because they were 'our birds', 'our garden friends', only passing through Europe on migration back to their true homes in the United Kingdom and the warm embrace of Britain's suburban back gardens. International action was needed, then, to prevent the extinction of our familiar garden avian visitors.

With its tone of moral outrage and its strong humanitarian attack on these 'Continental' behaviours, the press coverage drew on a century of campaigning by bird protection societies in Britain, Northern Europe, and North America that had sought to ban forms of hunting and trapping associated with the killing of wild birds for food. In Britain, the Seabirds Preservation Act (1869) had introduced a close season for the killing of thirty-three species of seabirds in the face of evidence of significant population declines caused by sport shooting, hunting for the millinery trade, and harvesting for food. The Wild Birds Protection Acts of the 1880s had extended this protection to most other species of birds, while the Protection of Birds Act (1954) put these laws on a national footing, giving legal protection to the majority of wild birds and their eggs in Britain. The 1954 Act explicitly prohibited a range of killing methods associated with the harvesting of wild birds for food. These included the use of snares, springs, traps, nets, and birdlime, together with the deployment of live bird decoys.[4] Since the early 1950s, the International Council for Bird Preservation (ICBP) had also campaigned against the killing of birds in Belgium and the Mediterranean basin, with its national sections providing evidence to their parent organization to help in the lobbying of national governments.

In the context of the rising influence of bird protection across Northern Europe and North America, UK legislation was notable, though far from unique, for the way it provided specific legal exemptions for the harvesting of certain wild birds for food. The

Seabirds Preservation Act (1869) had allowed the taking of young seabirds for this purpose, while the third schedule of the Protection of Birds Act (1954) listed a group of 'sporting or quasi-sporting birds' that could be killed, including for food, outside the close season. Most of these were wildfowl, but the list also included a number of wading birds and a few seabirds. The generally poorly enforced bird protection acts of the 1880s had also failed to effectively eliminate the trade in songbirds as food in the United Kingdom. During World War I, poulterers at Leadenhall Market in London and their customers – West End stores – were still selling blackbirds, song thrushes, mistle thrushes, redwing, and larks.[5]

If the cessation of the widespread harvesting of wild birds for food in Britain took some time to finally happen, its demise was hastened by economic change in the first half of the twentieth century as much as by law reform. Modern fishing methods, improvements in agriculture, the development of a food processing industry, and the creation of new forms of employment undermined both the importance of wild birds as food and the centrality of killing birds to the making of a living for sections of the population. As naturalist Frank Fraser Darling wryly suggested in 1947, 'American wheat, Argentinian beef and New Zealand mutton' had saved a number of British animals, including several of its bird species.[6] Similarly, for Julian Huxley, the introduction of kerosene (for lighting) and of tinned foods had encouraged some of the bird hunters of the North Atlantic to reduce their dependence on seabirds, allowing the bird populations to increase.[7]

As the concessions within the bird protection law acknowledged, however, the killing of wild birds for food did not entirely disappear. During the twentieth century the practice survived as a central part of the lives of two marginal communities within the United Kingdom: in the Western Isles of Scotland on the islands of St Kilda (until its evacuation in 1930) and Lewis. Both traditions of killing seabirds received legal protection, even if conservationists periodically contested this. The St Kildan and Lewisian hunters of seabirds formed part of a wider North Atlantic culture of seabird fowling that linked together communities in northwestern Scotland, the Faeroe Islands, and Iceland. All of these communities developed distinctive – if related – techniques for harvesting the prodigious natural resources of wild seabirds that were a central part of the ecology of their island lives.

In this chapter I explore the bird-harvesting cultures of the North Atlantic islands, bringing them together with the parallel traditions of bird harvesting practised within the Mediterranean basin and in parts of the mainland of Northern Europe. In doing so I explore the material and subjective attachments to wild birds found within bird-harvesting cultures. These were attachments produced through the techniques and devices for catching and killing birds and through the culinary traditions that made use of this natural resource.

In turn, the harvesting of wild birds played an important role in shaping local and regional identities. The latter were sharpened by the external criticisms that were made of the killing of wild birds throughout the twentieth century. In the case of the North Atlantic islands, defending the customs and traditions of seabird fowling formed part of a desire by these peripheral communities to protect their way of life against established centres of power, be that the political classes of the Central Belt and Westminster for the hunters of the Western Isles, or of Danish authority for the Faeroe Islanders and Iceland. Similarly, bird hunters in Belgium and southern Europe defended their customs against national governments or the increasing power of the European Economic Community (EEC) and the European Union (EU).

The issue of the harvesting of wild birds across different human communities posed a moral conundrum for bird conservationists. In Britain, conservationist critics were, as we will see, often sharply split in their respective judgements of the parallel cultures of bird harvesting, with 'continental barbarousness' contrasted with a more sympathetic, if romanticized, support for the island traditions of the North Atlantic. In this regard, one of the paradoxes of the protectionist response from the late nineteenth century and through the twentieth was the way it differentiated between 'legitimate customs' and 'uncivilized traditions' that threatened the populations of breeding and migratory birds. British conservationist responses were, in this sense, shot through with wider feelings for and attitudes towards the peripheral cultures of the North Atlantic and Southern European ways of life.

In its first section, the chapter documents the practices of bird harvesting and the attachments to wild birds found within the North Atlantic islands. At the heart of this account is the harvesting of young gannets, or guga in Gaelic, undertaken by the guga hunters of Ness on the isle of Lewis in the Western Isles. The Nessmen's hunting of the guga was subject to repeated disapprobation from bird conservationists in the first half of the twentieth century, but they and their allies proved particularly adept at winning legislative support and ultimately conservationist backing for the legal and cultural validity of the hunt.

In protecting their custom, the Nessmen benefitted from the sympathetic attitude towards the other 'bird people' of the Western Isles, the St Kildans, who had both fascinated and appalled government officials, social reformers, churchmen, and naturalists since the islanders' way of life had first been documented in the seventeenth century. The St Kildan's fashioned a society entirely dependent on the seabird harvest from their small, isolated archipelago forty miles west of the Outer Hebrides. While this was a way of life that ceased to exist following the evacuation of the islands in August 1930, St Kilda's bird-centred culture continued to fascinate commentators and to be remembered even as the island was reimagined as a conservation and military landscape once the human inhabitants had left.

While the Lewisian guga hunters came from a Gaelic-speaking, deeply religious community that was closely related to that of the St Kildans, both populations shared much with the bird fowlers of the other North Atlantic islands: the Faeroes and Iceland. These latter islands were subject to Danish rule for much of their history, and both were occupied by Britain from 1941 to 1944. The Faeroes became a self-governing community within the Kingdom of Denmark in 1948, and Iceland became an independent republic in 1944.[8] As on St Kilda and Lewis, the islanders of the Faeroes and Iceland had access to large seabird colonies, and these birds formed an important part of the diet of their communities. They both developed a set of techniques for hunting and eating the birds that closely paralleled those of their British compatriots, forging traditions of hunting, sharing, and preserving the seabirds that formed a near-common culture across the North Atlantic islands.

The chapter's second section moves on to explore the different traditions of harvesting wild birds within Southern Europe and in parts of mainland Northern Europe. These traditions involved not seabirds but migratory passerines or songbirds including thrushes, warblers, and larks that passed through mainland Europe in concentrated numbers during their autumn migration from Europe to Africa. Bulked up by fat accumulated to sustain them on their long journeys and moving in large flocks along traditional flyways, the birds were an attractive and welcome source of food for many human communities in the autumn. They formed part of the culinary traditions of the south of France, Cyprus, and Northern Italy in particular, with the annual bounty being celebrated within local festivals as much as traditional dishes. The pickled birds were also exported to expatriate communities in the United Kingdom and sold in high-class stores in London's West End into the 1970s.

The exporting of the birds to Britain and the scale of the killing in the Mediterranean basin provoked a sustained campaign by bird protection societies at various moments during the twentieth century. It became a flashpoint in the wider antagonism between the more pro-conservation countries of Northern Europe and the pro-hunting traditions of the South. This clash of sensibilities was increasingly informed by the new bird science and by studies of population and migration, with bodies such as the RSPB seeking to show that the harvesting of birds common to all European countries was unsustainable from the perspective of bird population numbers.

THE BIRD PEOPLE OF THE NORTH ATLANTIC ISLANDS

In the spring of 1939, Max Nicholson, James Fisher, and Julian Huxley – leading figures within the British Trust for Ornithology (BTO) and advocates of the new ornithology – landed on the remote Hebridean islands of St Kilda. They were there to conduct a three-day census of the islands' most important breeding seabirds. Reflecting on the trip later in life, Nicholson recalled that it was 'among the happiest and most dazzling of my life'.[9] All three naturalists were struck – as many had been before them and more have been since – by the stunning natural beauty and drama of the islands.[10] On the main island of Hirta, the soaring cliffs of Conachiar rose to more than 1,000 feet, with the islands and stacks of Boreray, Stac an Armin, Stac Lee, and Soay equally dramatic against the strong Atlantic currents that pounded the islands.

The census party was also stunned by the birdlife on the islands. Alongside its unique form of wren, St Kilda hosted one of the largest colonies of gannets in the North Atlantic and was home to the longest-established, and one of the biggest, colonies of fulmars in

Britain. It also held large populations of breeding auks, including puffins and guillemots.

The numbers of seabirds on and around the islands also brought back memories of the people who had lived on St Kilda, probably for more than a thousand years, and whose way of life was intimately connected with the birds.[11] Signs of their presence were everywhere on Hirta, the archipelago's main island, as the party surveyed the island's wildlife. This included the well-preserved houses of the main street of Hirta and the cleits, or stone drying chambers, which were scattered across the island. These had been used by the St Kildans to preserve and store food and hunting equipment. The ghosts of this population, who had left the islands just nine years before Nicholson and his party arrived, hung heavy in the warm St Kildan air that spring. Reflecting on the trip soon after his return, Huxley described the unique culture that had existed on the islands and the extraordinary community that had sustained a way of life that was built upon the harvesting of the seabirds. Huxley dubbed them the 'bird people' because of the centrality of the wild birds to the whole St Kildan way of life.[12] He was not, however, the first traveller to be riveted by the 'bird people' of the archipelago.

Since the first published account of the islands was written in 1698, the extraordinary human community of St Kilda had fascinated travellers to the islands. With a population that at its peak reached only 180 people, and which by the early twentieth century consisted of just over seventy individuals, St Kilda was both socially marginal to wider British society and geographically remote.[13] And yet it generated prodigious interest far beyond the actual size and sociological significance of the community – from travellers, from churchmen, from government officials, from tourists, and from readers of newspapers on the mainland. The latter were fascinated and appalled in equal measure by what Victorian and Edwardian social typologies typically referred to as the strange 'race' of St Kildan people.

What all visitors to, and commentators upon, the islands could not fail to spot, however, was the centrality of seabirds to their way of life. The islanders subsisted almost entirely on the flesh and eggs of the local seabirds.[14] Between March and November the killing of seabirds dominated their lives. Three species, abundant on the archipelago, were especially important: the puffin, the gannet, and the fulmar.

Of these three species, the people of St Kilda had their most intense and intimate relationship with fulmars. For much of the nineteenth century the islands were the only place in Britain where the birds nested. These small albatross-like birds were the islanders' most important source of food, feathers, and oil. The islanders had a practical knowledge of the birds' breeding habits, and they never took fulmar eggs, understanding that the females only ever laid one. The harvesting of the fulmar focused on the young birds and began each year in August. As Steel notes, the fulmar harvest was the 'busiest, most exciting and most important' event in the St Kildans' year. It represented – as writer Norman Heathcote suggested in 1900 – the islanders' only form of sport, and it cemented their connection to their island home.[15]

The harvesting involved men, women, and children, and it began on 12 August each year. Climbing in bare or stockinged feet, the fulmar fowlers tended to operate in pairs, with ropes tied around each other's waists, working areas of the cliffs allocated by the general assembly of adult men of Hirta.[16] The young birds were skilfully grabbed before they could regurgitate the foul-smelling oily vomit that they used as a defence. They were then strangled, and the oil was poured out from their inverted bodies into pouches slung around the fowlers' necks. The oil was a precious resource on St Kilda, used for lighting and bartered with the islands' landlord in lieu of rent. The number of birds killed was substantial. Steel estimated that when the population of the islands stood at 100 in the years between 1829 and 1843, an average of 12,000 fulmars were killed each year. This amounted to around 118 birds per inhabitant. The last harvest, in 1929, saw the killing of 4,000 birds, averaging out at 125 fulmars per inhabitant.[17]

In the nineteenth century, salt was used to preserve the birds, but this technique was combined with the longer-established method of air drying the fulmar flesh in the hundreds of stone-built, turf-roofed storehouses, or cleits, constructed on the island. The fulmars were typically eaten during the winter months, taken with potatoes as the main (lunchtime) meal of the day, having been boiled or stewed. The birds were supposed to have tasted like beef, and subsisting on the flesh of these birds, along with the islands' gannets and puffins, preserved the teeth of the islanders despite the absence of toothbrushes on St Kilda.[18]

The attitude of many outsiders to the St Kildans' prodigious slaughtering of seabirds reflected the broader fascination with this strange, apparently backward, group of people. It was generally

Figure 7.1 St Kilda fulmar harvest (early 1900s).

accepted as an essential part of their way of life, and the St Kildans' evident preference for the flesh of seabirds over other kinds of food, including beef and mutton, was acknowledged as part of their extraordinary culture. There is certainly little evidence of church missionaries, government officials, or even the islands' landlord seeking to convert the St Kildans to the food consumed by city dwellers in Glasgow or Edinburgh. The St Kildans may have taken to using salt to help preserve their food and acquired a liking for biscuits and sweets supplied by tourists, but official policy seems never to have been driven by a desire to change the islanders' food habits, despite widespread concern in the Scottish Office about the health and living conditions they faced. Furthermore, in the key bird protection legislation of the late Victorian period – the Seabirds Preservation Act (1869) and the Wild Birds Protection Act (1880) – it was acknowledged in law that 'certain birds required for the support of the Inhabitants of the Island [of St Kilda]' were exempted from the general protection.[19]

Later commentators on the seabird fowling traditions of the islanders writing after the evacuation of the islands in 1930 have reinforced this sympathetic attitude. It was particularly noticeable in the attention given to the St Kildans by leading 'new ornithologists'

Figure 7.2 Bird fowling on St Kilda (early 1900s).

in the middle decades of the twentieth century. James Fisher, in his monograph on the fulmar (which used the census data gathered during his 1939 visit to the islands), celebrated what he called the 'brave human culture based on birds' that had existed on St Kilda. He described how the evacuation had brought to an end the 'last great seabird fowling community of the North Atlantic', and he told of how fulmars had flourished on St Kilda while this civilization had existed. The birds had been 'exploited carefully, traditionally' by the islanders, he claimed, and the birds' numbers had increased.[20] In his subsequent study of the *Sea-birds of the North Atlantic*, co-authored with Ronald Lockley, Fisher had further suggested that the St Kildans, together with the other seabird fowling communities of the Faeroes and Iceland, were 'unconscious conservators', harvesting the birds to ensure the long-term viability of the bird populations.[21] Kenneth Williamson, senior research officer at the BTO and a visitor to the islands in the late 1950s, similarly celebrated the 'great personal integrity' of the St Kildans. He directed his ire at those commentators who had presented only a negative view of the islanders and focused on their poverty, squalor, and disease. The islands were better remembered, he urged, as 'the showplace … of birds and great cliffs and cragsmen without equal in the British scene'. For Williamson, if the culture of St Kilda had declined in its latter years it was because it had been, in his emotive phrase, 'corrupted' by its contact with the outside world.[22]

Williamson, like Fisher, Lockley, and Huxley, was part of that group of 'island lovers' who played key roles in shaping the 'new natural history' of the inter-war and war years. All saw St Kilda as an 'island laboratory' where contained communities of animals, plants, and humans could be studied.[23] As Huxley put it, here was 'a laboratory on the doorstep' for British biologists – one in which they could find 'an inexhaustible store of material for the study of evolution in action'.[24]

Like Huxley, Fisher, and Lockley, Williamson was particularly drawn to the rugged islands of the North Atlantic. These included Fair Isle, part of the Shetland Isles, where he acted as the director of the Bird Observatory from 1948 to 1957. He also had an intense interest in and affection for the Faeroe Islands, an archipelago situated between Shetland and Iceland. Williamson had been stationed there as part of the British Army's Headquarters Faeroe Islands Force during the British occupation between 1941 and 1944.

He used his time there not only to document the islands' bird life and indulge in birdwatching, but also to marry one of the islanders, Esther Rein, and to produce a detailed ethnological study of the 'life and scene' of the Faeroes. The resulting book, *The Atlantic Islands,* described the Faeroese community, its cultivation of the land, and the fishing, whale hunting, and bird fowling that they practised. Williamson detailed the hunting of the *juglabjorg,* or 'bird mountains', on the north and west coasts of the islands. The islanders, he noted, had developed distinctive hunting techniques for each species, and these had evolved into specific fowling traditions. Puffins were the most important bird for the food economy of the Faeroes, and the average summer harvest on the island of Mykines consisted of about 30,000 birds.[25] The birds were caught with a large triangular net fastened to a long pole, called a *fleygastrong,* as they flew around the cliffs. Fisher and Lockley, in a nod to English schools' sports, compared the *fleygastrong* to a '3 foot lacrosse net'.[26] Williamson was riveted by the skill of those involved in using the *fleygastrong,* describing it as requiring 'a high standard of coordination of the hand and eye, strength of arm and wrists and a well-nigh perfect sense of balance'. It also looked to him like great sport: a 'fascinating game'.[27]

Alongside the physical strength and skill used in hunting puffin, Williamson described the accumulated knowledge of the birds held by the islanders and how they selectively killed only non-breeding adults. Birds flying into the cliffs with food in their beaks for young – know as *sildeni* or 'herring bearers' – were spared, thus allowing the puffin colony to reproduce itself.[28] While the islanders harvested large numbers of gannets, guillemots, and fulmars, puffins were the most valuable food item. The majority was salted down for the winter, with a typical household of five in the late 1940s preserving around 200 birds, the puffins stored under the rafters of the islanders' houses.[29] During his stay on the island, Williamson was given puffin for lunch on several occasions, and like Fisher and Lockley he was enthusiastic about its taste. 'I found it rich and tasty,' he concluded, and a 'lundi [puffin] lunch is certainly one to write home about.'[30]

While the Faeroe Islanders evidently shared Williamson's liking for puffin, they were much less keen on the taste of fulmar. On a visit in the late 1930s, Ronald Lockley suggested that while they ate the young fulmars with barley bread for breakfast, the islanders disliked the flavour of the adult birds, describing it as 'dry and tasteless, like paper'. They were also not well disposed to fulmar in general as they felt the birds could drive out puffins from their nesting sites. If, as Lockley colourfully recounted, the locals had to boil fulmar-soiled clothes in cows' urine to remove the stench, he felt that they viewed the poor little lundi as a likely victim of this scourge (the fulmar and its vile vomit).[31]

These differing attitudes towards particular seabirds and the distinct culinary traditions associated with them separated the people of the Faeroes from the bird fowling culture of St Kilda. Like the bird fowlers of the Westmann Islands (part of Iceland), the Faeroe Islanders harvested and ate seabirds as part of a mixed peasant economy in which agriculture and fishing were also key elements. This also marked them out from the St Kildans, who were exclusively dependent on birds for their sustenance.

Across the bird fowling communities of the North Atlantic, then, as sympathetic observers like Fisher, Lockley, and Williamson noted, there developed similar, if distinct, techniques and devices for taking the birds, as well as different traditions of preserving and cooking them. All these new ornithlogists saw the 'bird people' as 'heroes and heroines', as Lockley put it when referring to the Faeroe Islanders. The greatest threat to their ways of life came from outside interference and the lure of modern industry, communications, and leisure. Lockley was so affronted at the incursion of elements of modern life into the inter-war society of the Westmann Islands that he complained to the British vice consul. As he put it in his letter, what he had hoped would be (referring to the town of Kaupstadur) a 'simple Icelandic fishing village … in which the people lived as simply, as primitively as the St Kildans used to' in fact looked no different from a 'thriving English fishing port'. The town had been transformed by electricity, central heating, the telephone, and radio. Modern fashion had also remade the habits of young women. 'As I walked the narrow streets,' he lamented, 'I saw … that lipstick, plucked eyebrows and silk stockings had come this far north.' This contrasted with his earlier encounter with young women in traditional dress collecting milk on the Faeroe Islands. In a moment of erotic epiphany, he had been struck by their beauty and charm. Because of the encroachment of modern life, many young islanders in the North Atlantic would, he feared, prefer to experience the life they saw in the movies 'rather than experience the hard adventures of the bird cliff'.[32]

THE GUGA HUNTERS OF LEWIS

Identification with the island life of the people of St Kilda, the Faeroes, and Iceland – as well as an enthusiasm for their bird-centred cultures – shaped how new ornithologists like Lockley, Fisher, and others viewed another peripheral coastal community within the United Kingdom. These were the Gaelic-speaking crofters of the Port of Ness on the Isle of Lewis in the Western Isles – the last place where seabird harvesting survived in the United Kingdom through the mid to late twentieth century (and into the present). The economy of the Port of Ness had been built on fishing from the latter part of the nineteenth century when it became a busy fishing port. By 1900, more than 12,500 hundredweight of ling was caught there annually.[33] Farming sheep supplemented the fishing and, as Donald S. Murray recalled from growing up in Ness in the 1950s and 1960s, these two forms of food were supplemented by a third: guga, or young gannets.[34]

Since at least the sixteenth century, groups of fisherman from the port had undertaken a late summer harvest of the young offspring of the gannets that nested on the rocky island of Sula Sgeir, forty-five miles west of the Butt of Lewis. Two boats of nine or ten men from Port of Ness had traditionally competed to complete the annual harvest of young gannets. By the mid to late twentieth century, this had been reduced to one boat.

The hunt was subject to a degree of secrecy and the Nessmen were generally reluctant to discuss their seabird harvesting with outsiders. One of those who was able to gain the trust of the Nessmen was the police photographer and naturalist James MacGeoch. He visited Sula Sgeir with them in 1954, 1957, and 1958, bringing back an account of their harvesting techniques and of the social life of the guga hunters.[35] MacGeoch described how the crews sailed to the island on or around 29 August, remaining there for between two and three weeks. During their stay the men camped in stone bothies and survived on water and provisions they had transported with them, as well as eating some of the young gannets they caught.

The harvesting was an intense business, with an average of just under 2,500 birds taken each year (between 1919 and 1958). The hard work of killing the gannets went on daily, except for Sundays when the strict sabbatarianism practised by the Nessmen required them to cease working. Sundays were spent reading the bible and relaxing. The hunting itself was a highly organized, collective effort. The young gannets, specifically those that were nearly fully fledged, were snared with long catching poles by the neck, clubbed on the head, and then passed by the skilled catchers in a human chain up to the top of the island. Here, the birds were decapitated, plucked, their feathers singed off over a peat fire, and their wings removed, and then the birds were split open, their rib cages and entrails drawn out. Other men worked to salt the birds and to fold them into a square shape ready to be assembled into what MacGeoch called a large beehive-shaped pile or what John Beatty later referred to 'a symbolic wheel of meat'.[36] When the hunt was completed for the year, this pile of guga was transported to the boats and taken back to Port of Ness. The villagers typically gathered around the harbour walls as the men returned, creating a sense of excitement and a festival atmosphere. The birds were divided up between the crew and then sold to the wider community. The villagers stored the majority of the birds for use during special celebrations or family events. A proportion found their way to the Hebridean diaspora on mainland Scotland and in North America.

Sefton Gordon, a resident of the Isle of Skye, described the pickled guga as having the strong smell of mackerel and looking 'more like pickled fish than birds', though he conceded they were considered 'excellent eating'.[37] Later commentators suggested that the preserved birds had 'the texture of good steak and the taste of kipper… It was neither fish nor fowl, but somewhere in-between.'[38] The guga were traditionally boiled, with the water changed regularly and excess fat drained off. This cooking process, a legacy of the years when only a minority of Scottish islanders owned ovens, had few enthusiasts, generating as it did a pungent smell that prompted many to choke or gag. The boiled guga was served with boiled potatoes. As food tastes changed in the Western Isles from the 1960s, younger Lewisians turned away from a food they described as being 'similar to duck stewed in cod liver oil and salt [and having] the consistency and flavor of oiled chamois leather with a frame of bone'. Even those that liked it claimed it gave the eater a prodigious thirst.[39]

The place of guga harvesting and guga eating within the economy, customs, and culinary practices of the people of the Port of Ness was not beyond criticism, however. In the late 1930s an article in the journal *British Birds* unwittingly sparked an intense controversy about the harvesting – or slaughtering – of gannets on Sula Sgeir. Inspired by the emergence of population studies being promoted by the BTO, Malcolm Stewart spent the summer of 1937 making a census count of the gannetries of Sula Stack and Sula Sgeir.[40] From the relative comfort of his father's chartered steam yacht, Stewart visited both islands and sought to calculate the relative increase or decrease in the gannet populations since an earlier count in 1932. Estimating the population in terms of breeding pairs, he suggested that there had been a significant decrease in the birds' numbers since the early 1930s. Reflecting on the possible causes for this decline, Stewart focused in on the large number of nestlings that were taken each year from Sula Sgeir by the Nessmen. With his personal antipathy to this practice surfacing within his scientific paper, Stewart scoffed at the 'supposed food value' of the birds and suggested that the men from Ness seemed to have 'nothing better to do than to undertake the unpleasant voyage to Sula Sgeir'. 'Not even the most grumbling of Lewisman,' he went on, 'can complain of a food shortage, and this annual venture is nothing short of an unnecessary destruction of bird life... Small wonder that this gannetry ... is gradually being exterminated.'[41]

Stewart's article prompted leading figures within the RSPB and the Scottish Society for the Protection of Wild Birds (SSPWB) to write to Ross and Cromarty county council and then to *The Times* to protest against the killing of young gannets. Citing Stewart's assertion about the decline of gannets on Sula Sgeir 'due to [the] continual slaughter', Montagu Sharpe, chairman of the RSPB council, and John M. Crosthwaite, honorary secretary of the SSPWB, urged the local county council to outlaw the killing and to give Sula Sgeir the status of a bird sanctuary. Ross and Cromarty council's response was to deny the request and argue that the harvest was an essential part of the livelihoods of the people of Ness. The Nessmen themselves spoke up for the importance of the harvest. Seeking to defend their custom, the experienced hunter Norman Maclean argued that there was no cruelty involved in their killing of the birds and that the gannets formed an important part of the economy of the district. Any prohibition of the harvest would seriously affect the living standards of the people of Ness.[42]

The local newspaper, the *Stornaway Gazette and West Coast Advertiser*, took up the Nessmen's case. It robustly challenged what it saw as the condescending view of the tastes and customs of the people of Ness expressed by Sharpe and Crosthwaite, adding that the RSPB was more than happy to condone the killing of English game for sport while apparently finding the humane harvesting of gannets 'incomprehensible'. The paper emphasized the virtues of the people of Ness and the centrality of the guga to their material and emotional well-being. As it assertively put it: 'There is no more honest, upright or law-abiding community in the Empire, and their deep Christian piety finds its expression in the holding of family worship even in the isolation of Sulisger [Sula Sgeir]... Who will say that the people of Ness will not be one whit better [off] mentally, morally or spiritually when ... they obtain all the necessities of life at their doors from a grocery van.'[43]

While he was concerned about the fate of the gannets, Frank Fraser Darling, author of the New Naturalist study *Natural History in The Highlands and Islands*, was drawn to defend the hunt in similar terms to the Nessmen's local paper. As Fraser Darling put it in a letter to a colleague in the spring of 1938, and after confessing to some equivocation, 'I must say ... that I would rather see the people wisely handling a natural food resource gained by their own labour than buying the same amount of fat and protein in the form of margarine and bully beef.'[44]

In defending the moral character of the Nessmen and the importance of the hunting of the guga as a custom central to their way life, Fraser Darling and the *Stornaway Gazette* articulated a support for the hunt that found resonances among its other defenders. Across the letters pages of *The Times* through the late summer, autumn, and winter of 1938 and into early 1939, supporters of the seabird fowling practices of the Lewisians emphasized again and again the courage and endurance of hardship of the Nessmen, drawing unfavourable comparisons with the easy and pointless sport of game shooting.[45] They also emphasized the essential part played by the hunt in the local economy. Defenders of the hunt further contested the claim of bird protectionists that the gannets were declining on Sula Sgeir and challenged whether there was in general any risk to their population in the Western Isles.[46]

One of the most eloquent defences of the Sula Sgeir hunt came from Ronald Lockley. He argued that all the evidence suggested that gannet numbers were increasing across the North Atlantic, including in their North of Scotland colonies. Interfering with the livelihoods of the Ness 'fisherman crofters' was thus unjustifiable in conservation terms. He praised the courage and stoicism of the Nessmen in undertaking the demanding journey to Sula Sgeir, and he made a final observation about the humanitarian criticism of the hunt. As he put it: 'On humane grounds this taking of the young gannets is no more objectionable than the modern methods of killing farm poultry or animals, and this taking is a good deal more humane than the shooting of pheasants, grouse, etc.'[47]

While the debate in *The Times* fizzled out by early 1939, it remained a live issue for many of the protagonists through 1940 and into the war years. In February 1940 Malcolm Stewart wrote to James Fisher, the secretary of the BTO, summarizing anonymous testimonies he had collected from those opposed to the Sula Sgeir hunt and the responses of the Nessmen to their critics. He prefaced his precis of the evidence with a reiteration of his opposition to the 'slaughter'. 'Quite honestly,' he explained to Fisher, 'the Nessmen have no real need to go on with the fowling except for old times sake. They rather enjoy their annual outing and have made it into a saga. In my opinion, however, ancient custom is not necessarily of its own accord good.'[48]

The testimonies collated by Stewart repeated the arguments about whether the hunt was really necessary to the local economy of Port of Ness, about whether the uncontrolled slaughter had already led to serious population declines of seabirds, and about whether the clubbing to death of the guga was crude. In reply the Nessmen argued that the killing of the gannets was no more cruel than the way pheasants were shot or domestic chickens had their necks wrung. They again insisted that the guga added much to their livelihoods and, above all, that the hunt was an 'immemorial custom'.[49]

There is no record of James Fisher's reply to the evidence provided by Malcolm Stewart, but in a paper on the breeding distribution, history, and population of the North Atlantic gannet published in 1944, Fisher and his co-author H.G. Vevers suggested that, contrary to Stewart's argument, full protection of the birds was neither necessary nor desirable given the value of the custom to the people of Ness.[50] Fisher and Vevers did, however, recommend that future harvesting should be 'consciously planned' and limited to no more than 1,000 birds.[51] This was a more considered response to Stewart's evidence and arguments than that of Tom Harrisson. Corresponding with Fisher about the gannet census and his own figures, Harrisson attacked Stewart's integrity: '[He] has no idea of ornithological method... I ought to know, having been with him at St Kilda and North Rona, as well as at school. But he was always a great "bullshitter".'[52]

The arguments made by 'new naturalist' supporters like James Fisher and Ronald Lockley reinforced the appeal to time-honoured custom and tradition as a central plank in the defence of the Sula Sgeir hunt. This 'customary consciousness' was invoked to claim the right to practise the hunt in a way that was similar to the invocation of 'custom' of trade made, as E.P. Thompson has argued, by many groups that were subject from the eighteenth century to processes of economic rationalization and innovation, such as enclosure and new work disciplines.[53] This conservative, oppositional idea of custom also *did* something, as Thompson again argued. In the case of the Nessmen and their supporters, it reinforced their sense of community against the forces of modernization and change, including bird conservation. The claims to tradition, custom, and habit formed part of the defence – particularly for the community of Port of Ness itself – of a Gaelic-speaking, Puritan crofting identity: an identity suppressed since the post-Culloden moves by the

Hanoverian State and its successors to destroy and later marginalize the Gaelic language and the Highland way of life associated with it. Eating guga, as Donald S. Murray has argued, was a way for the community of Port of Ness to bind itself both to its locality and to the practices of its ancestors. It was, in Thompson's terms, part of a 'rebellious' conservative culture.[54] For the new naturalist 'island lovers' like Lockley, Fisher, and Fraser Darling, the defence of the hunt was also about supporting a brave, tough, and romantic way of life and the intense attachments to and knowledge of seabirds made by these bird-centred cultures.

The force of this appeal to custom – alongside the evidence from bird census studies that the gannet populations of the British Isles were generally thriving – had an important influence on the special provision made in 1955 to give legal exemption to the hunters of Sula Sgeir within the Protection of Birds Act (1954). While the passing of the Act had initially failed to protect the hunt, an amendment was swiftly introduced in 1955 following an appeal made by Ross and Cromarty county council to the Secretary of State for Scotland. This allowed the gannets of Sula Sgeir to be moved to the third schedule of the Act. This formed the Wilds Birds (Gannets on Sula Sgeir) Act (1955).[55] This amendment to the law was later bolstered in a revision to the Protection of Birds Act undertaken in the late 1960s. Revealing the degree to which bird protection societies like the RSPB were now sympathetic to the cause of the Nessmen, in moving the second reading of the Protection of Birds Bill in the House of Lords in May 1965 Lord Hurcombe, president of the Society, spoke in defence of the guga hunters of Sula Sgeir: 'For at least 400 years,' he argued, 'teams of crofter-fisherman ... have made annual visits in late summer to this isolated rocky island for winter food... This practice has in no way endangered the survival of the gannetry and, if continued, will not do so.'[56]

HUNTING AND CULINARY TRADITIONS IN SOUTHERN AND NORTHERN EUROPE

The accommodation with the residual culture of guga hunting by bird conservationists in the United Kingdom, together with the sympathetic attitudes of leading 'new naturalists' to the seabird fowling cultures of the North Atlantic (both past and present), contrasted sharply with a much more critical view of the widespread practice of harvesting songbirds around the Mediterranean basin and in parts of mainland Northern Europe. Certainly the leading bird conservation organizations of the post-war period – the RSPB and the ICBP – both became increasingly concerned about the effects on the migratory birds that either summered or wintered in Europe from their trapping and hunting for food as well as for sport.

The main hunting and trapping areas corresponded to the principal migratory bottlenecks through which many millions of songbirds (and other species) passed during their spring and autumn migration. The migration was particularly concentrated during the autumn, and the traditional flyways were well known to the human communities through which the birds passed. These included flyways through the Pyrenees and the Alps and across the islands of the Mediterranean, particularly Cyprus, Malta, and Sardinia.[57] Hunting and trapping also had a strong regional presence in Lombardy and Veneto in Northern Italy; in Southern France; in the Spanish provinces of Valencia, Aragon, and the region of Catalonia; and in Belgium.[58] Among the most important bird species hunted for food across these regions were skylarks, thrushes (mostly song thrushes and redwings), some finch species, and warblers (including blackcaps).

A range of similar techniques for trapping the birds was used across Europe. In Southern France, skylarks and thrushes were caught with snares, stone crush traps, and lime sticks. Cage traps were also used to capture ortolan bunting, one of the French culinary tradition's most prized species. The trapped birds were fed in cages to fatten them up for eating.[59] In Northern Italy bird trappers used a device called a *roccoli* to catch (principally) thrushes. The *roccoli* were towers surrounded by nets to which migratory birds were lured.[60] In Spain, a device called either a *barracia* (in Aragon and Valencia) or a *paramy* (in Catalonia) was used to catch mainly thrushes. It consisted of a grove of trees pruned so that sticks covered in lime could be inserted along the trees' branches. Whistles, caged birds, or (later) recordings of birdsong were used to attract overflying migrants to drop in and rest in the grove. Upon alighting on the branches, the birds became stuck to the lime, a viscous and sticky substance extracted from plants. The lime prevented the birds from flying and they dropped to the ground where the hunters collected them.[61] Lime was not only a highly destructive substance for the birds, covering and ripping off their feathers, but also indiscriminate, with any species of bird entering the *paramy* being fatally contaminated. Birds of prey, owls, and many kinds of songbirds were often killed alongside the target species.[62] In southern France and Belgium, long lines of nets were used within coastal dunes and farmland to catch songbirds, including skylarks. A hunter in a hide would drop the nets on the bird lured towards them by live decoys.

The birds caught around the Mediterranean basin and in parts of Northern Europe had historically formed an important source of sustenance: an autumn bounty for rural communities. The birds were incorporated into distinctive culinary traditions and ways of eating. In France, the breast meat of skylarks, known as larks' tongues in English, was eaten, but the most celebrated gourmet meal made from small birds was 'ortolan'. This consisted of a plate of plump ortolan bunting bodies in a bowl with a napkin over them. The gourmet diner placed the napkin over their heads, took in the aroma, and placed each small bird whole into their mouths. Described as being like 'lumps of fat 3 ounces in weight', the whole bird was crushed in the mouth to release an explosion of flavour.[63] Until the hunting of ortolan bunting was made illegal, around 50,000 birds were caught each year in France.[64]

In Italy, with its strong tradition of regional recipes and products, home cooking, and the survival of small-scale producers, a range of distinctive wild bird dishes flourished.[65] This remained the case even as growing prosperity changed rural areas after World War II. The fresh meat of these birds was an attractive alternative to the predominantly cured and preserved food that dominated the diets of those in the poorer, rural parts of the country. As Capatti and Montanari suggest, outside of the elites, food – be that cheese, vegetables, or meat – was 'flavored, monotonously, with salt'.[66] Four main species of songbird or passerine were regularly eaten. The first was skylark, or *allodola.* This was typically used to flavour ragus, and in Tuscany it was wrapped in pancetta and grilled on skewers between chunks of bread and bay leaves. Sparrows, passeri, and warblers, especially blackcaps, formed the basis of a dish called *beccafico,* meaning 'figpecker' and referring to the habit of autumn migrant songbirds to feed on ripe grapes and figs. This was seen to endow their flesh with a sweet flavour. The birds were tightly packed in a tray and baked with pine nuts, breadcrumbs, raisins, and the juice of oranges.

Thrushes, or *tordo,* were cooked on skewers in a similar way to the *boccafico.* Into the twenty-first century, 10,000 birds were still being trapped in Lombarby and Veneto. In more recent years, however, the *tordo* have been replaced with the more socially acceptable alternative of farmed quails. The annual thrush harvest, however, was particularly important in Tuscany, including in the town of Montalcino, where an annual festival called Sagra de Tordo took place each autumn.[67] In the Brescia region, thrushes and other small birds formed a central part of the celebrated dish *polenta e osie* (oatmeal with birds).[68]

The culinary traditions and the songbird hunting that supported them became an increasing target for bird conservation organizations from the early 1950s. For the RSPB this represented the return to a campaign that the Society had waged during World War I. Then, however, the Society had been most concerned about the eating of largely British-sourced songbirds. The RSPB's secretary Miss Gardiner, in a campaign that mirrored the Society's attack on the plumage trade, had targeted stores in London's West End and encouraged poulterers not to sell 'English wild birds' to them. Gardiner was suspicious of the claim made by the stores that their birds came from abroad, and she used interviews with shop staff to challenge this defence. In a letter to *The Times*, Lady Geraldine Mayo supported Gardiner's assertion about the origins of the birds that were for sale, suggesting that 'many dozens of blackbirds, thrushes … larks and redwings' sold in the stores came from Norfolk. It was important to stamp out what she called 'the wholesale murder of our native songbirds' as they added so much 'charm' to life in Britain and were an ally in the 'war against mankind's greatest enemy, the insects'.[69]

While the RSPB's campaign against the sale of songbirds as food was overtaken by other campaigning priorities, the problem of continental songbird harvesting was taken up by the ICBP after World War II. In 1950 the first international agreement protecting wild birds in the post-war period, the International Convention for the Protection of Birds, was signed in Paris. It condemned and sought to outlaw the practice of bird liming. The British section of the ICBP received complaints that bird liming was being undertaken in Cyprus and raised the matter with the ICBP's central committee.[70] One of those to highlight the problem was David Bannerman, the well-known ornithologist and former curator of the British Museum of Natural History. In his book *Birds of Cyprus* he described how literally thousands of blackcaps found their way each year on Cyprus 'to the purveyors of beccaficos through the barbarous medium of the lime-sticks'.[71]

The ICBP also revealed the existence of bird trapping in other European countries. The Belgian section of the ICBP reported that the country's Minister of Agriculture had agreed to extend the netting period for wild birds from 1 October–15 November to 20 September–27 November. Much of this legal trapping was undertaken in the Ardennes region of the country and involved migrating finches, buntings, and thrushes. The majority of the birds caught and killed found there way into local shops. The ICBP, including its British section, was angered not just at the extension of the hunting season, but also by its continued existence *tout court*. As it argued in its letter to the government: 'The birds in question [are] protected in other countries, they [are] not the property of any one country, some of the species were valuable insect-destroyers, and catching the birds en masse was a barbaric practice.'[72]

In 1958 the British section of the ICBP focused in on what its annual report called the 'mass destruction of birds in Italy'. It suggested that the trapping of songbirds in their hundreds of thousands as they migrated through the country was causing consternation to bird conservationists, particularly within Northern Europe. The Northern European protectionists were disturbed because it was, they felt, chiefly 'their summer birds that pay the highest toll for this annual slaughter'. The British section encouraged tourists to Italy not to visit restaurants where dishes like *boccafico* were served and to avoid Italian towns where 'Feasts of birds' took place during the late summer or autumn.

These views were transmitted in a letter written by the RSPB's Lord Hurcombe (chairman of the British section of the ICBP) to the Italian Minister of Agriculture, in which he set out the argument that 'British tourists [felt disgust] at the use of songbirds as food'.[73] Not all British visitors to the Mediterranean necessarily supported this view, however. In his autobiographical account of the three years that he spent living on Cyprus in the mid-1950s, the author Lawrence Durrell was more matter of fact about the consumption of what he called *boccafico*. Describing eating the pickled birds in the convivial setting of a friend's tavern, he recalled that 'together we crunched the small birds to bits as we tasted the wines of Cyprus and sagely assessed them'.[74]

Through the mid to late 1960s, protesting against the mass destruction of migratory birds in Europe became a major activity of the ICBP as a whole. Their focus was particularly on the killing that took place in Belgium, southwest France, and Italy, and the national sections in those countries particularly affected led the organization's response.[75] The ICBP took great heart from the publicity in the British press against this hunting, particularly the high-profile story run by the *Sunday Mirror* about the 'massacre' of songbirds in Belgium.

The RSPB had also become heavily involved in this campaign, collaborating with the ICBP and later working with the press to publicize the threats to songbirds posed by continental trapping and hunting.[76] In the autumn of 1969 the Society sent their technical officer (investigations), Richard Porter, at the request of the ICBP, to southwest France to give his support to a campaign led by the French bird protection society to stop bird trapping in the region. In a photo essay published in the March–April 1970 edition of the RSPB members' magazine *Birds*, Porter documented the 11,000 trapping stations that operated in the department of Les Landes. His essay included photographs of the long clap-nets being used in dunes or across fields, with images of skylarks trapped and being killed by the hunters. Porter noted how many of the restaurants in the region served larks and other songbirds, with the market in Bordeaux selling species like skylark for 15 francs per dozen and song thrushes for 3 francs each. Other parts of the harvest were sent to Paris or, after being turned into pâté, made their way to West End London restaurants.[77]

Two years after Richard Porter's visit, another member of the Society's head office, David Lea, participated in protests against bird trapping in southwest France led by the Ligue Francais pour la Protection des Oiseaux (LFPO). This followed a large demonstration in Belgium against the practice. Lea was involved in leafleting nearby towns in the region and in direct action in the form of breaking up the nets used by the hunters. The scale of the netting, covering long stretches of the coast between Biarritz and Gironde, shocked Lea, and the scale of destruction wrought upon migrating birds during his visit appalled him.[78]

The RSPB was also active in protesting against bird trapping in Belgium and Cyprus, making representations to both the Belgian and Cypriot ambassadors in London.[79] An article in *Birds* detailed the scale of the killing in Cyprus, suggesting that as many as 7 million migrating birds were killed on the island each year. The professional trapping was concentrated in the villages of the southeastern part of the island, where lime sticks were used to catch the birds. Conceding that this practice had been going on for centuries and that it had historically had little impact on breeding populations, the article warned that more recent threats to wild birds from toxic chemicals, land drainage, and reclamation and industrial development combined to make the traditional practices on Cyprus no longer justifiable. Invoking the idea of a common, Western European birdscape, the article asked: 'Can Palearctic species continue to withstand the annual slaughter of their numbers in Cyprus?'[80]

In July 1973 the RSPB called for a ban on importing songbirds from Cyprus, while in October 1975 the *Sunday Times Magazine* ran a special report on the killing of songbirds in Belgium, France, and Italy under the headline 'Europe's birds run gauntlet of death', with the sub-heading 'Slaughter'.[81] Photographs showed the traps, birds caught by their legs on lime twigs, and the result of the killing in the form of wild bird pâté. Repeating the proprietorial approach of the *Sunday Mirror*'s 1969 exposé, the *Sunday Times Magazine* article drew out the implications of this practice for its readers: 'Summer songbirds from your garden are now dying in their millions. For them the great winter migration south ends not.' Using images of the hunters in France and Italy 'displaying the day's carnage' of killed birds, as the reporter put it, the article drew particular attention to the plight of skylarks, highlighting their place within the English landscape. 'They arrive in their millions [in southwest France], skylarks that sang all summer over English downs ... but [with] over 10 million [other birds] they will never arrive' in their wintering grounds.[82] For huge numbers of these migrating birds, the report solemnly revealed, 'this English spring will be their last'.

The *Sunday Times* report had a big impact in the United Kingdom, and the RSPB received nearly £3,000 from readers of the article to assist it in stopping the killing. Some of this money was given by the Society to an Italian bird protection group to launch a campaign to introduce new bird laws in the Italian parliament. The issue of protecting migratory birds in mainland Europe also dominated the 11th European Conference of the ICBP in West Germany in 1976. The conference established a European Committee to tackle the problem of the mass killing of migratory birds.[83]

During 1978 and 1979, Siegfried Woldhek, from the Dutch arm of the European Committee for the Mass Destruction of Migratory

Birds, researched the problem of bird hunting in the Mediterranean basin, talking to policymakers, police, hunters, and conservationists. His report, 'Bird killing in the Mediterranean', pulled no punches in documenting the scale of the hunting. As its introduction suggested, 'everywhere in the Mediterranean, guns, air-rifles, lime-sticks, nets and snares are ready to take a bird's life'.[84] Woldhek estimated that around 15 per cent of the probable 1 billion birds that migrated through the Mediterranean were shot or trapped each year. Bert Axell, the pioneering warden of the RSPB's Minsmere reserve and later the Society's land use adviser, recalled the explosive impact of the report, describing it as 'one of the most effective, thought-provoking documents ever written in our business'.[85]

Woldhek's report for the ICBP and the Council's wider campaign was successful in building on and helping to consolidate moves within the EEC to establish common rules across the Community for the protection of threatened bird populations. In December 1977 all the environment ministers of the nine member states of the EEC agreed on the new protection measures.[86] These formed the basis of the EEC's Birds Directive, established in 1979, which gave extensive protection to migratory and resident European birds. The Directive of the Protection of Wild Birds (1979) banned the killing or trapping of native European birds across the territories of its member states. It did, however, exempt a number bird species from this law, including twenty-five species of wildfowl, twenty-two wader species, five species of dove, and twelve passerines.[87] Article 9 of the Directive, and the later principle of derogation of European laws, additionally allowed some trapping and hunting, including for 'scientific' purposes, to continue in (notably) France and Italy.[88] This meant that while new laws restricted the hunting of songbirds, the harvesting of the skies over Europe during the autumn continued in the era of new European bird conservation laws.

CONCLUSION

The continued hunting of songbirds (and other species) across parts of southern Europe revealed the deep-rooted nature of these practices, the political support that the hunting lobby was able to mobilize, and the broader complex politics of bird hunting and its relationship to conservation. For the supporters of bird trapping and hunting, the practice was and remains a sustainable way of harvesting a natural resource that had little effect on bird populations. Rather than emerging from a position of hostility and wanton cruelty towards birdlife, as its critics suggested, harvesting wild birds instead revealed, for its defenders, a deep ecological knowledge of the birds, their habitats, and their movements. It was borne of a proximity to these wild birds and a connection with and understanding of them. Attacks on their bird killing and eating habits from national governments, foreign conservationists, and bodies such as the EEC/EU simply served, in many ways, to strengthen the pull of these practices and the 'customary consciousness' associated with them.

Resisting external pressure for change in the name of custom and tradition has certainly deepened some of the divisions between hunters and conservationists. This has occurred at the same time as changes in the social make-up of hunters in countries such as Spain and Malta have meant that it is affluent and mobile hunters who most staunchly defend the traditional practices.[89] Supporters of hunting in the Mediterranean basin have also wondered whether it forms an essential part of a 'Mediterranean' identity rather than a 'European' one, and so fits into the idea of there being a 'circum-Mediterranean' region conceptualized as a distinct 'culture area'.[90] For sympathetic academic commentators such as Mark-Anthony Falzon, hunting wild birds is centrally rooted in its own intense feelings or passion for those birds – a passion organized through the knowledge and devices required to lure the birds to prepared landscapes and involving sporting expertise as well as a visceral pleasure in acts of killing. Bird trappers and bird hunters, then, have ordered their relationships to wild birds within managed sporting and harvesting landscapes, using technologies like nets, traps, and guns and the power of human hands.

For conservationists, the supporters of trapping and hunting failed to recognize the fact that the migratory birds that are killed were part of the 'natural heritage' of all Europeans and not the property of any one interest group. Furthermore, in challenging the hunting of songbirds, groups like the ICBP and the RSPB, as we have seen, suggested that hunting could not be justified because it put additional pressure on bird populations at a time when they were already threatened by toxic chemicals and habitat loss. For bird protectionists, whatever its ethical justification in the past as the harvesting of a natural surplus, data on population declines of the birds that were regularly hunted meant that traditional hunting could no longer be justified.[91]

The mobilization of bird census data by international bird conservation groups offers a clue as to why there was an apparent double standard in the views of bodies like the RSPB to continental songbird killing and the guga hunters of Sula Sgeir. For all the controversy provoked by the latter tradition in the 1930s and 1940s, the fact that new ornithologists such as Ronald Lockley and James Fisher could show that the population of gannets in northern Scotland was not declining allowed them, and latterly the RSPB, to support the guga hunt. In giving the hunters of Port of Ness legal recognition of their hunt and cultural validation of their custom, conservationists and naturalists were able to reconcile themselves with the killing. As we saw, this support was also informed by the identification made by some of the new ornithologists with the bird-centred cultures of the North Atlantic. The deeply embedded place of seabirds within these peripheral communities spoke to the love of seabirds felt by Lockley, Fisher, Huxley, and others.

There was also, as we saw, a powerful pull towards these dramatic Atlantic islands for the new ornithologists, and the lure of the heroic fantasy of island living. While this may have been coloured by a romantic conceit – one that wanted to fix in time the ways of life of the North Atlantic peoples – it was also motivated by a sense that these communities were 'unconscious conservators'. In other words, in their closeness to, knowledge of, and dependence on wild birds, the people of St Kilda, Lewis, the Faeroes, and Iceland were motivated to protect wild birds even as they killed them for food. As such, they trod more lightly on the earth than did those in more developed parts of Europe, notwithstanding the support for bird conservation in these societies.

There is one final dimension of the close identification with the seabird fowling communities that emerged within the new ornithology that is worth noting. This concerned the way in which the techniques for catching seabirds used by the hunters – including the nets, the traps, and the rock climbing methods – were appropriated by bird ringers from the 1930s. In a telling photograph from his book on the Atlantic Islands, Kenneth Williamson showed a bird ringer descending down a cliff attached to a rope to ring young gannets. It echoed the bird fowling techniques of the North Atlantic peoples and revealed an enduring connection between bird harvesting and bird conservation in some of the devices and techniques used to get close to wild birds. As Williamson himself described it, bird ringing, particularly when it involved 'some energetic climbing', also incorporated the sporting element of bird fowling. It was, like harvesting them, 'a splendid mental and physical exercise' as much as a useful technique in the new bird science.[92]

Conclusion

In November 1944 the US Fish and Wildlife Service issued a press release that revealed the answer to a long-standing puzzle of bird migration.[1] Through the work of its bird-banding programme, led by Frederick C. Lincoln, the Service had discovered the wintering grounds of the chimney swift, a common summer avian visitor to North America. While the winter range of almost every other North American bird had long been known, the chimney swift had kept its secret up until this point. Following intensive banding or ringing studies undertaken mostly by amateur ornithologists since 1936 (in which nearly 375,000 swifts had been banded), thirteen bands taken from the dead bodies of swifts had been returned to the Service from Peru, via the American embassy in Lima. Indigenous Peruvian hunters in the River Yanayaco area of the country had killed the birds and used the return address stamped on the bands to help convey them back to the United States. Through this circuitous route, new knowledge had been generated about a hitherto-mysterious aspect of the life history of a common North American bird.[2]

The content of the press release pricked the attention of its author, Rachel Carson, who was at that time working for the US Fish and Wildlife Service. She had published her first book, *Under the Sea-Wind*, three years earlier and was seeking to establish herself as a full-time writer. Carson had been collecting material for a biographical sketch of the chimney swift for some time, and the news about the birds' wintering range offered her an opportunity to fashion a timely article for a magazine like *Reader's Digest*.

Carson's notes played with her own memories of watching chimney swifts flying low over a freshly cut meadow close to her home. The birds had been drawn down from the upper reaches of the sky by rain and by the lure of the insects disturbed by the mowing. The encounter had become fixed in her memory, fusing an intense moment of joy: the freshly cut meadow, the fine droplets of rain, the sweet smell of the grass, and the flight of the swifts.[3]

The article, which Carson initially titled 'Ace of nature's aviators' but later retitled 'Sky dwellers', did not in the end include her recollection of the moment of reverie with the swifts over the meadow.[4] It did, though, mix the findings of the new bird science with Carson's characteristic depiction of the 'wonder' of the natural world. She emphasized the aerial prowess of the swifts and their extraordinary lives lived out exclusively on the wing, noting how the birds only landed at night to roost in dead trees or chimneys.

Even more striking than their airborne lives was the swifts' apparently mysterious disappearance in the autumn. Gathering in huge numbers along the Gulf coast of America only to then suddenly disappear, Carson suggested that it was as though 'the skies had swallowed them [the swifts]'. As she explained, however, drawing on the press release, the mystery of the birds vanishing had now been explained by the efforts of bird banders and the recovery and return of their bands by the Peruvian hunters. The birds did not – as naturalists in America had thought well into the twentieth century – hibernate in swamps before emerging in the spring. Rather, as the science of bird banding had been able to show, they migrated to South America.[5]

Carson's article not only registered the increasingly influential role played by the new bird science within popular natural history, but also showed how this was interfused with strong emotional investments in the lives of birds by those drawn to observe and study them. The saga of the chimney swift additionally revealed how the 'new ornithology' enjoyed a continuing relationship with bird hunting and harvesting practices. Not only did banders deploy nets and traps similar to those used by hunters in catching birds to ring, but they also collaborated with those who pursued wild birds for sport and for food to facilitate the return of the bands. The life history of the chimney swift further showed how people's relationships with the bird was sustained across different landscapes and environments, with the birds brought into close encounters with different human populations in geographically distant locations. Across these diverse spaces of encounter the birds themselves lived out cosmopolitan lives as they moved along traditional flyways that connected their summering and wintering grounds.

These themes – the mobilizing of amateur enthusiasm within the new bird science; the emotional attachments to wild birds; the ties between natural history and sporting and harvesting practices; and the different relationships forged with wild birds by distinct human communities across geographical space – have all been central to the account developed in this book. Bringing together the new forms of bird conservation and recreational birdwatching with field sports and bird harvesting traditions has been central to the book's claims. It has allowed an interrogation of the productive relationships formed between these practices through the central decades of the twentieth century. This has been at the heart of the book's counter narrative to established histories that have tended to segregate field sports, natural history, and conservation.

The book has shown how these different bird-centred practices were in a sustained dialogue with each other through the short twentieth century. This dialogue was generated at an institutional level by the organizations that represented the various outdoor pursuits and by the statutory bodies that oversaw nature conservation and popular recreations. The dialogue between the different interest groups was most developed in relation to the conservation of wildfowl. As we saw, it involved both international agreements between national governments and the work of bodies like the US Fish and Wildlife Service and Nature Conservancy in the United Kingdom. On both sides of the Atlantic, these state agencies acted as brokers between the interests of bird conservationists and wildfowlers.

In the United Kingdom, the close and productive ties formed between WAGBI, the RSPB, and the Wildfowl Trust developed these relations. Similarly, organized falconry and bird conservation in the United Kingdom found ways of working together from the 1930s on, with the 'toxins crisis' of the late 1950s and 1960s helping to further consolidate their common interests. Even in relation to the more divisive issue of harvesting seabirds for food, the supporters of the new bird science, bird conservationists, and the bird hunters managed to find common ground and balance the imperatives of conservation with the defence of selective hunting.

If the demands of legitimate sport, customary bird harvesting, and modern bird conservation could find forms of accommodation with each other during key decades of the twentieth century, this study has also revealed surprising connections in how these different bird-centred practices established subjective and material attachments to wild birds. The first revolved around the historical associations and common roots of the technologies and techniques for bringing birds closer to people across these practices. One striking example that we saw concerned the manipulation of landscapes within bird conservation and wildfowling. The latter practice bequeathed to bird conservation techniques for managing protected spaces for wild birds, with bird reserves influenced by the way sporting landscapes were shaped. In addition, the bird hides and the reed screens used to shield footpaths to hides on these reserves were adapted from the screens used in duck decoys and from the shooting pits and screens used by wildfowlers. Moreover, sporting landscapes continued in the twentieth century to shape encounters between wildfowlers and wildfowl that were remarkably similar to those produced on bird reserves. If each practice ultimately styled human–avian relations in different ways, with wildfowling organized around the imperative of killing and the technology of the gun, what they held in common in terms of landscape practices was equally notable.

Similar connections were also evident in the relations between bird harvesters and bird protectionists. In his pioneering approach to bird ringing, Heinrich Gatke learnt from and adapted the knowledge and technologies of Heligoland's bird catchers to establish devices and techniques for modern bird study. Gatke took the simple traps used by the islanders in their harvesting of avian migrants and developed his more elaborate Heligoland trap to catch, ring, and release the same migrating birds. This established a connection between bird catching and bird ringing that was sustained through the subsequent development of bird ringing and banding programmes in Britain, Western Europe, and North America. Nets, tape-lures, traps, and ropes moved from the practices of the bird harvesters to those of the bird ringers. Part of the undoubted squeamishness of some bird protectionists in the early years of the century to the practice of bird ringing stemmed from the close parallels between catching birds for food and bird ringing. Birds trapped in bird ringers' nets looked remarkably like those caught by bird harvesters.

If similar technologies and devices connected the different bird-centred practices, they were also drawn together by closely associated feelings and passions for birds. This has been the second central claim of my book.

One key dimension of these feelings concerned the competitive, sporting elements that linked these bird-centred practices. This element was evidently central to sporting culture. Fowlers like James Wentworth Day thrilled at the challenge of pursuing elusive and wary prey. Peter Scott similarly recalled the adrenalin rush of punt gunning as a young man and the consummation of that pursuit with the moment of the kill. Like wildfowlers, falconers were enlivened by the drama of the hunt and by its risks and dangers (including the chance of losing a much-loved falcon).

These passions were not just integral to sporting culture, however. For the new post-war birdwatchers, particularly the 'tally-listers', sporting emotions were evident in their enthusiasm for 'bagging' rare birds and in the lengths they would go to in order to add to their collection of life ticks. For the more committed tally-listers, the rarity became not the occasional delight of birdwatching, but the principal goal of the hobby. For this group of birdwatchers, the pursuit of rare and scarce birds was central to a sporting tournament of value that defined their recreation.

However, while they might have been at the extreme end of their hobby, tally listers were far from unique among the new birdwatchers in wanting to experience a quickening of the pulse. In developing the principles of the new ornithology, Max Nicholson had wondered, as we saw, if birdwatching was not the 'most scientific of sports or the most sporting of sciences?'[6] For Desmond Nethersole Thompson, the thrill of finding nests and outwitting birds remained at the heart of his study of and feeling for wild birds. Roger Tory Peterson also confessed to the stirring of his sporting juices as he sought out birds across America and beyond. Even the more casual birdwatchers who processed along a screened track in the summer of 1959 to view the Loch Garten opsreys from the RSPB's carefully managed viewing hide, or the ones who were drawn to the Society's Suffolk reserves to watch its avocets at close quarters, were motivated by the excitement of seeing rare birds and the competitive privilege it conferred upon them.

Sporting elements were also present among the hard necessity that shaped the harvesting cultures of the North Atlantic world. For some observers of St Kilda's fulmar harvest, the work of killing and collecting birds and their eggs was certainly made more enjoyable for its participants by the way it undeniably combined forms of sporting adventure. Kenneth Williamson, James Fisher, and Ronald Lockley could also not help but see a sporting dimension in the puffin and gannet catching undertaken by the Faeroe

Islanders. As Williamson confessed, climbing the cliffs to get to young gannets was great fun as well as being the perfect mental and physical exercise.

If conservation-led birdwatching, field sports, and the traditions of bird harvesting shared common sporting passions, they were also joined by their feelings for the avian landscapes they cohabitated during a key period of the twentieth century. As RSPB secretary Peter Conder acknowledged, birdwatchers and wildfowlers were drawn to the same wild marshes and fenland in their pursuit of wildfowl. Putting in many hours, often at antisocial times of the day, both outdoor recreations were united by a common purpose: to experience the spectacle of flocks of geese and ducks in dramatic landscapes. As we saw in the case of Peter Scott, these experiences with wild geese constituted a sonic as much as a visual geography, with the birds' calls as integral to their appeal as the sight of skeins of geese against dawn skies over brooding marshland. Scott's personal journey from wildfowler to bird conservationist was undoubtedly eased by the continuities between the experience of 'avian landscapes' that were integral to both sets of practices.

Falconers also shared with birdwatchers the pull to wild landscapes, including the Scottish uplands and the uncultivated lowlands of southern England. Birds of prey helped to animate and enchant these sporting landscapes for falconers. For all these pastimes, charismatic birds, including birds of prey, were read as an integral part of the landscape, itself frequently seen through the prism of national heritage and national nature. If this reading of the landscape was sometimes connected to the proto-Medievalist tendencies of British falconry, it fitted with the anti-modern impulses of falconry culture and the flight – for at least some of its practitioners – from urban modernity. The author T.H. White was an exemplar of this desire to find solace in a sport that seemed to offer qualities that were missing from contemporary life. White, like Wentworth Day, however – another sportsman with antediluvian impulses – mixed a reactionary flight from modern life with an enthusiasm for speed, fast cars, and modern means of communication.

This mixture of modern and anti-modern sensibility was also present among some of the leading 'new ornithologists'. They too combined an enthusiasm for science and the new technologies of bird study with deep feelings for the rugged islands of Britain's west and northwesterly coasts. These were landscapes on the edge of modern society: isolated Edens that offered solace and escape from aspects of the contemporary world.

A somewhat different, but related, emotional geography was at work among the bird hunters of both the North Atlantic and the Mediterranean. For these communities, the rituals associated with the killing and consumption of wild birds bound them closely to the places in which they lived. This connection to place and to a local identity was expressed, in the case of the Mediterranean hunters, through their knowledge of the seasonal bounty of migrating birds, the traditional devices of trapping used, and the food dishes and festivals with which this bounty were associated. Similarly, for Lewisians, the eating of guga was an affirmation of their rebellious, regional identity. Many of these harvested species of bird may have lived mobile, cosmopolitan lives, but the relationships that human populations had with them helped to fix their hunters to the landscapes and places through which the birds passed or to which they returned to breed.

Wild birds could also play companionable and human-liked roles within field sports and conservation. Maternal and pedagogic impulses were central to the emotional connection that falconer T.H. White established with his birds. In the tragic tale of his goshawk 'Gos', White played out various projections and identifications in relation to the bird, seeing it as a child to be mothered and as an unruly schoolboy to be educated and trained. In similar terms, wildfowler-turned-conservationist Peter Scott formed a parent-like relationship with one of his wild geese, whom he named 'Annabel'. As we saw, Annabel was Scott's 'goose child', invested with a unique avian personality and freighted with Scott's paternal/maternal feelings towards it. In his later conservation studies of wild swans at Slimbridge, Scott continued to see wild birds through an individualizing and familial lens. Identifying birds by their bill patterns, he gave them human names and established the familial connections and family trees of their dynasties. These swans were perhaps less his 'swan children' and more families that mirrored his own. His studies of them involved Scott's own his wife and children as co-researchers in the recording and studying of the birds. In a more quotidian, domestic setting, backyard and back garden birdwatchers – that core constituency of organizations like the National Association of Audubon Societies and the RSPB – were not only drawn to the beauty and colour of their garden birds, but saw them as sources of companionship. For these birdwatchers, common garden birds brought solace and comfort as familiar friends, with the birds, we might suggest, being almost as knowable as the people in their lives.

The sense of companionship and the fierce intensity of feeling expressed in human–avian relations across field sports and the new birdwatching were not the only feelings shared across these outdoor recreations. The dramatic declines in the populations of birds of prey and the encounters with dead and dying songbirds stirred emotional distress and feelings of loss. These were felt most acutely by birdwatchers such as those suburban American women recorded by Rachel Carson who found dead and dying birds in their gardens and neighborhoods following spraying campaigns or whose bird feeders were suddenly devoid of the colour and activity of birds. A sense of loss was also experienced by the American raptor watchers such as Charles Broley and Maurice Broun, who were both struck by the very visible decline of species such as the bald eagle. Birdwatchers on both sides of the Atlantic shared this sense of something essential being subtracted from the environment by the rapid decline of bird populations. These were feelings that falconers also experienced as their relationships with species like the peregrine falcon – the much-loved mainstay of their sport – were broken by the population crash of the birds in both Britain and America. As J.A. Baker powerfully expressed it, in the age of rapid environmental change and degradation, to look at the landscape and its populations of wild birds – be that for falconers or birdwatchers – was to see the shadow of death.

While wild birds could provoke shared feelings of loss, they could also stir deep visceral emotions. As we saw, both birdwatchers and falconers were drawn to the aerial prowess and the killing power of birds of prey. This enthusiasm for the hunting skills of these top predators drew falconers and birdwatchers into the violent, bloody aspects of the lives of birds of prey. J.A. Baker was again a key voice, articulating wider shared attitudes. Baker could find solace and redemption in seeking to become one with the falcons he watched, transfixed in part by the intensity of their power as killers. In this he shared with a generation of post-war falconers a pleasure in watching the acts of killing performed by birds of prey. Baker was not unique among birdwatchers either. A broader constituency of post-war birdwatchers, including the RSPB's mass membership, was also encouraged to thrill at the power and killing prowess of birds of prey through the way in which the birds were represented in birdwatching and countryside books such as the AA's *Book of the British Countryside*. Even within conservation-led forms of watching birds, this enthusiasm for the visceral aspects of how birds of prey hunted and killed was a live and active dimension of human–avian relations.

These violent and bloody dimensions of the appeal of birds of prey that linked the new birdwatchers and falconers were not strong enough, however, to prevent fractures appearing in the relations between organized falconry and bird conservation. Critics of the sport, such as the RSPB's secretary Peter Conder, worried that it was linked to bird crime through the illegal taking of young wild falcons, and, more fundamentally, that it was a practice that limited the natural modes of life of birds. Similar divisions also became apparent through the 1970s in the relations between wildfowling and bird protection. Growing ethical differences disturbed the shared identifications and feelings with birds across conservation, sporting, and harvesting practices. The centrality of killing to field sports became increasingly incommensurable for many with the styling of human–avian relations within conservation-led birdwatching.

Deeper divisions also appeared between bird protectionists and the Mediterranean bird hunters. For organizations such as the ICBP and the RSPB, these (largely) southern European traditions represented unacceptable forms of killing. Not only were they appalled at the methods used by these bird harvesters, but they also saw the annual slaughtering of birds as putting already-threatened bird populations under greater pressure. Using population studies, bodies such as the ICBP and the RSPB sought to delegitimize established harvesting customs and traditions.

In their attacks on the customs of the bird harvesters, bird conservationists increasingly moved beyond the largely national mapping of bird populations and ideas of 'national nature' that had shaped bird protection and bird science in the first half of the twentieth century. This built on the pioneering studies of wildfowl conducted in the early decades of the century, including Frederick C. Lincoln's idea of 'flyways'. By the 1970s the debate about the killing of migratory songbirds in Europe increasingly worked to understand birds' mobile lives and the transnational spaces they inhabited. The international networks of leading birdwatchers, conservationists, and sportsmen supported studies of bird migration that revealed the strikingly cosmopolitan lives of many birds and how they transcended attempts to fix them to any particular national territory or national nature. Wild birds shared their lives with different human communities along traditional flyways and in their wintering and breeding grounds.[7] Through studies of migration and the efforts of international conservation, bird protectionists and birdwatchers increasingly sought to understand wild birds across interconnected transnational spaces and to protect wild birds and their habitats through international cooperation.

These forms of 'democratic science' and their cartographies of birds' cosmopolitan lives have remained central to international bird conservation. Their legacy is evident in three major reports published between 2013 and 2019 by fifty-three conservation organizations into the state of Britain's wild birds (and its other animal and plant communities). Titled 'State of nature', these reports captured the dramatic declines in many of the bird populations found in Britain and its overseas territories. As the 2019 report concluded, the total number of breeding birds in the United Kingdom had declined by 44 million from 1967 to 2009.[8] Across these reports, many wild birds, some of which were once common, were shown to be in steep decline. There was a poignant irony in these figures. For all the success of bird conservation organizations in making visible the threats to wild birds, in strengthening conservation law internationally, and in growing their own memberships, they have been unable to halt the relentless pressure from socially driven environmental change and the apparently inexorable slide of once-abundant wild birds towards scarcity and, for some, extinction.

Conflicts between the different bird-centred recreations have sharpened in this era of big declines in wild bird numbers and the wider environmental crisis. The successful lobbying by Mediterranean hunters of their national governments and the EU has meant that a number of species of European birds continue to be heavily hunted along their migration routes between Northern Europe and West and Sub-Saharan Africa. This includes the illegal hunting of birds of prey in Cyprus, Malta, and southern Italy. Bird conservationists across Europe have mobilized against this continued killing, with the lobbying of national and supranational governments supported by direct action.

In the United Kingdom, birds of prey are also at the heart of intense conflicts between sporting and conservation interests. In a recent RSPB investigation unit report titled 'Bird crime', the Society documented 550 incidents of birds of prey having been poisoned, shot, or trapped and killed over a five-year period (to 2017).[9] The RSPB's investigations have led it to directly confront countryside and sporting interests, including the British Association of Shooting and Conservation (BASC), the successor to WAGBI. The BASC have continued to present themselves as a pro-conservation body and the field sports they defend as an integral part of the rural economy. But in a break with the close relationship with conservationists in the post-war decades, the BASC is more aggressively critical of groups like the RSPB.[10] The result has been to entrench the ethical and political divisions between conservation-minded birdwatchers and the bird-centred field sports. The dialogical relations between these hobbies that were so productive for much of the twentieth century have given way to more incommensurable feelings for birds.

Overlaying these new divisions has been the continuing growth of interest in watching, feeding, and recording wild birds among urban and suburban populations in North America, Britain, and parts of Western Europe. Shaped by a welfare and conservationist ethic and much more distanced from sporting and hunting cultures than was previously the case, wild birds have become objects of pleasure and solace for more and more people, offering them connection to a natural environment that is itself seen as an increasingly precious resource for human well-being and flourishing. This includes the 47.8 million Americans who watch birds and the million British people who took part in the RSPB's annual Big Garden Birdwatch in 2021.[11] As Rachel Carson noted in the wake of the controversy around her book *Silent Spring*, the joy of close encounters with familiar birds and the sight of birds in flight were essential parts of an enlarged sense of being human – a sense rooted, for her, in an ecological understanding of the natural environment and an assertion of the value of 'more-than-human worlds'. Relations of this kind with wild birds and the landscapes and places that they help to animate remain central to the new birdwatchers of the twenty-first century.

This book has sought to place feelings and emotions at the heart of its account of the styling of human–avian relations, not just within birdwatching but also across some of the other ways in which people encountered wild birds through the short twentieth century within field sports and bird harvesting cultures. Its challenge to historians, sociologists, and animal geographers is to attend more fully to these subjective dimensions of the ways in which people purposely seek out encounters with wild birds and other non-human animals. This means developing the study and theorization of emotions not just in relation to the ordering of human social relations, but as a central element of the study of natural relations as well.[12] *Passions for Birds* has begun to open up this line of enquiry, and it encourages others to take forward the exploration of human emotions towards non-human animals as part of any account of human entanglements with our many lively companion species.

Notes

INTRODUCTION

1 Hall, 'For Allon White: Metaphors of Transformations', 1–25; Stallybrass and White, *The Politics and Poetics of Transgression.*
2 Tory Peterson, *Birds Over America*, V.
3 Berger, 'Why Look at Animals?', 3–30; Thomas, *Man and the Natural World*; Fudge, 'Two Ethics: Killing Animals in the Past and Present', in The Animal Studies Group, *Killing Animals*; Almeroth-Williams, *City of Beasts*; Benton, *Natural Relations*; Philo and Wilbert, *Animal Spaces, Beastly Places*; Guha, *Environmentalism, A Global History*; Landry, *The Invention of the Countryside*; Matless, *In the Nature of Landscape.*
4 Dunlop, *In the Field Among the Feathered*; Barrow Jr, *A Passion for Birds*; Barrow Jr, *Nature's Ghosts*; Bircham, *A History of Ornithology.* See also Mynott, *Birdscapes;* Moss, *A Bird in the Bush.*
5 Sheail, *An Environmental History of Twentieth Century Britain*, especially chapter 5; Sheail, *Nature in Trust*; Sheail, *Nature Conservation in Britain.* See also Evans, *A History of Nature Conservation in Britain*; Mabey, *Common Ground*; Cocker, *Our Place.*
6 Griffin, *Blood Sport*, xii; Hoyle (ed.), *Our Hunting Fathers*; Marchington, *The History of Wildfowling.*
7 Mackenzie, *The Empire of Nature*; Adams, *Against Extinction.*
8 Mackenzie, *The Empire of Nature*, 202.
9 Warren, *The Hunter's Game*, 2–25.
10 Philo and Wilbert, *Animal Spaces, Beastly Places*; Nagai, Jones, Landry, Mattfeld, Rooney, and Sleigh, *Cosmopolitan Animals*; Wilson, *Seeking Refuge*; Barrow, *Natures' Ghosts*, 315–17.
11 Matless, *Landscape and Englishness*; Matless, *In the Nature of Landscape*; Matless, 'Versions of Animal–Human, Broadland, c. 1945–70'; Matless, 'Moral Geographies of English Landscape'; Matless, 'Action and Noise Over a Hundred Years: The Making of a Nature Region'. See also, inter alia, Brace, 'A Pleasure Ground for the Noisy Herds?'; Lambert, *Contested Mountains.*

12 Matless, *Landscape and Englishess,* 254.
13 Haraway, *When Species Meet.*
14 See Wilson, *Seeking Refuge*; Barrow Jr, *Natures' Ghosts,* 315–17.
15 Ducks Unlimited was formed in 1935. Its membership grew from 25, 000 in 1956 to 50, 000 in 1970 and 250, 000 by 1980 (www.ducks.org); NAFA, formed 1961, had around 2, 000 members by 1980 (www.n-a-f-a.org).
16 McKibbin, *Classes and Cultures,* 42–3; Cannadine, *The Decline and Fall of the British Aristocracy,* 637–51; Savage, *Identities and Social Change in Britain*; Mort, *Capital Affairs,* 57–63.

CHAPTER ONE

1 Nicholson papers, 'Great Crested Grebe Survey 1931–6', British Trust for Ornithology (BTO) Archives RB1; 'Great Crested Grebe 1935', AB1.
2 Harrisson and Hollom, 'The Great Crested Grebe Enquiry, Part 1', 62–92.
3 'Birds of the Meres', *The Times,* 4 October 1932, 11. The enquiry was subsequently repeated in 1935 in order to understand the effects of two very dry summers in 1933 and 1934 on the bird's population.
4 In the next chapter I track its influence on bird protection organizations and associations and look at the increasingly mass participation culture of birdwatching in the post-war years.
5 Dunlop, *In the Field Among the Feathered;* Barrow Jr, *A Passion for Birds;* Barrow Jr, *Nature's Ghosts*; Bircham, *A History of Ornithology*; Sheail, *Nature Conservation in Britain*; Evans, *A History of Nature Conservation in Britain*; Matless, *Landscape and Englishness;* Macdonald, 'What Makes You a Scientist Is the Way You Look at Things', 53–77; Toogood, 'Modern Observations', 348–57; Bircham, *A History of Ornithology.*
6 Matless, *Landscape and Englishness;* Macdonald, 'What Makes You a Scientist Is the Way You Look at Things', 53–77; Toogood, 'Modern Observations', 348–57; Bircham, *A History of Ornithology.*
7 Birkhead, *The Most Perfect Thing,* 13.
8 Matless, 'Visual Culture and Geographical Citizenship in England in the 1940s', 424–39; Davis, 'Britain an Island Again'; Davis, 'Militarised Natural History Tales', 226–32.
9 Barrow Jr, *A Passion for Birds,* 75, 154.
10 Nicholson, *The Study of Birds,* 57.
11 Matless, *Landscape and Englishness,* 307–8.

12 Nicholson, *The Art of Bird Watching*, 14–17, 148.

13 Ibid., 43, 158.

14 Letter from W.B. Alexander to E.M. Nicholson, 6 February 1934, summarizing Meinertzhagen's ideas, E.M. Nicholson BTO Archives, Correspondence, R4; Letter from Tom Harrison to James Fisher, 18 April 944, James Fisher BTO Archives, RB3. An adaptation of the use of aircraft was employed in the 1940s when the extensive wartime practice flights of RAF coastal command were used to photograph the seabird colonies on remote islands like St Kilda, Sula Sgeir, and Aisla Craig: see Nicholson, 'The Trust – Origins and Early Days', 25.

15 Nicholson, *The Art of Bird Watching*, 29–30.

16 Ibid., 97.

17 Ibid., 167; Nicholson, 'The Trust – Origins and Early Days', 19.

18 Nicholson, *The Art of Bird Watching*, 177.

19 *The Auk*, VXLIX (January 1932): 119–20.

20 Redington, *The Auk* (April 1931): 111; Burrow, *Passion for Birds*, 154.

21 Burrow, *Passion for Birds*, 168–71.

22 Ibid., 179; 'Griscom on Problems of Field Identification', *The Auk*, 53, 2 (April 1936): 238–40.

23 Schei, 'Ninety Years and Growing', Foreword, 20.

24 Nicholson, 'The Trust – Origins and Early Days', 18.

25 Macdonald, 'What Makes You a Scientist Is the Way You Look at Things', 56; Toogood, 'Modern Observations', 348; Hinton, *The Mass Observers*, 5.

26 Hinton, *The Mass Observers*, 4; Nicholson, 'Tom Harrisson', 596–7.

27 Tom Harrisson to James Fisher, 18 April 1944, James Fisher BTO Archives, Fisher RB3.

28 Huxley, 'Foreword'; Madge and Harrisson, *Mass Observation*, 6.

29 Madge and Harrisson, *Mass Observation*, 41.

30 'Starling Inquiry', 1941, James Fisher BTO Archives, R4.

31 Hinton, *The Mass Observers*, 5.

32 Lack, 'The Relationship of the Edward Grey Institute', E.M. Nicholson BTO Archives Correspondence, Tinbergen R4.

33 Reginald Moureau cited in Johnson, *The Ibis: Journal of the British Ornithologists' Union*, 19; Tinbergen, 'Recent British Contributions to Scientific Ornithology', 130.

34 Letter from E.M. Nicholson to Dr Kember, 11 December 1933, E.M. Nicholson BTO Archives, RB3.

35 Oxford Bird Census, Sub-committee Report, Note of a meeting held on 15 May 1931, E.M. Nicholson BTO Archives RB3.

36 *The Times*, 1 July 1933, 11.
37 Letter from E.M. Nicholson to Mr Dawson, 26 June 1933, E.M. Nicholson BTO Archives RB3.
38 BTO, *Annual Report*, 1949, 33.
39 H.F. Witherby, *British Birds* 26 (June 1932): 18. Ludlow Griscom made a similar defence of museum study in the 'new era' of field studies in the United States in 1936. He argued that skins continued to be important and supportive of the new 'field study'. Griscom was, however, critical of egg collectors, wishing that oologists would 'play the game' and give birds, especially rare birds, a chance. *The Auk*, VLIII (April 1936): 240.
40 Nicholson, *Birds in England*, vii.
41 Ibid., 249–51, 256.
42 Cole, 'Blown Out', 18–28.
43 Ibid., 20.
44 Ibid., 21–2.
45 Kearton and Kearton, *Our Rarer British Breeding Birds*, vi–vii.
46 Kearton and Kearton, *British Birds' Nests*. Their preface to the second edition of this book also took care to promote responsible nest and egg finding and challenged the collectors of rare breeding birds: see Kearton and Kearton, *British Birds' Nests, Revised and Enlarged Edition*, vi–viii.
47 *The Times*, 18 July 1923, 13.
48 During 1934 and 1935, the sports newspaper *The Field* also hosted a long-running exchange of letters on the ethics of egg collecting. When its editor Eric Parker published these as a book, *The Auk* reviewed the controversy: *The Auk*, VLIII (April 1936): 467.
49 B.H. Ryves, 'Egg Collectors', *The Times*, 13 March 1936, 10.
50 *The Times*, 20 March 1937, 8.
51 Dent, 'Saving British Birds', *The Times*, 26 November 1937, 13.
52 BTO, *Annual Report*, Spring 1936, 16.
53 The BTO did not finally condemn egg collecting until January 1954 when an Extraordinary General Meeting passed a resolution stating that the BTO considered egg collecting to be without scientific justification and a hindrance to field research in ornithology: see *Bird Study*, 1, 32. This position was reiterated in 1959: see *Bird Study*, 6, 1, 33.
54 Cole, 'Blown Out', 23.
55 Nicholson, 'The Trust – Origins and Early Years', 16.
56 'Bird Ringing', *Bird News and Notes*, Spring 1930, 121.

57 Established by Julian Huxley and James Fisher as the Hatching and Fledgling Enquiry in 1939, it was renamed the Nest Record Scheme: see BTO, *The Nest Record Scheme*, H. Mayer-Gross, Field Guide No 12, 1970.
58 Huxley, 'A Census of Nests', *Bird News and Notes,* Autumn 1930, 201.
59 Editorial reply to Huxley, *Bird News and Notes,* Autumn 1930, 201.
60 Perry, 'A Creed for Birdwatching', 214.
61 Ibid., 214.
62 Letters to the Editor, *The Field,* 9 March 1940, 379; *The Field,* 16 March 1940, 420.
63 The book was republished in 1973 based largely on the second edition of 1943.
64 Perry, *At the Turn of the Tide,* xiii.
65 Tom Harrisson's review of 'Lundy Isle of Puffins' by Richard Perry, Mass Observation Report No. 534, 30 December 1940.
66 P.H.T. Hartley, Letter to the Editor, *The Field,* 6 April 1940, 547.
67 Marren, *The New Naturalists,* 23–4, 39.
68 Ibid., 18, 25.
69 Ibid., 69.
70 Ibid., 182.
71 Fisher, Editorial, *The New Naturalist: A Journal of British Natural History* (Spring 1948), 3.
72 Ibid., 4.
73 Matless, *Landscape and Englishness,* 258; Macdonald, 'What Makes You a Scientist Is the Way You Look at Things', 56; Toogood, 'Modern Observations', 351.
74 Davis, 'Britain an Island Again', 1–3.
75 Fisher, Preface, *Watching Birds,* n.p.
76 Ibid., n.p.
77 Fisher, Preface to 1946 printing, n.p.
78 In my copy of the first edition of *Watching Birds,* the BTO application form has been cut out and was presumably used.
79 Fisher, *Bird Recognition,* volume 1.
80 Fisher, *The Fulmar,* 1.
81 Ibid., 147.
82 Ibid., 246.
83 Ibid., 141.
84 See chapter 7 for a fuller discussion of Fisher's views on the bird people of St Kilda.
85 Fisher, *The Fulmar,* 144.

86 Ibid., 2–6.
87 Letter from R.M. Lockley to John Buxton, cited in Niemann, *Birds in a Cage*, 48–9.
88 Lockley, 'Bird Migration Stations in Britain', *New Naturalist* (Autumn 1948): 165–71; Lockley, *Letter from Skokholm*, 23–40; Fisher, 'St Kilda, A Natural Experiment', 90–108.
89 Nicolson, in Lockley, *Letter from Skokholm*, 16.
90 Letters from Fisher to Lockley, 8 June 1944, 21 March 1949, James Fisher BTO Archives, RB3.
91 Niemann, *Birds in a Cage*, 8.
92 James Fisher to R.M. Lockley, 8 June 1944, James Fisher BTO Archives RB3.
93 Niemann, *Birds in a Cage*, 147.
94 Buxton, 'Birds of the Eichstatt District 2 September 1942–14 April 1945', Edward John Mawby Buxton (1912–1989) Papers, Alexander Library Archive and Manuscript Collection, Bodleian Libraries, University of Oxford, Box 3, 12, Notebooks; J.H. Barrett, E.J.M. Buxton, P.J. Conder, and G. Waterston, 'Migration of Birds at Warburg, 1942', Buxton Papers.
95 Buxton, 'Birds of the Eichstatt District', 1–2.
96 Editors preface, Buxton, *The Redstart*, xii.
97 Buxton, *The Redstart*, 3.
98 Ibid., 45.
99 Niemann, *Birds in a Cage*, 58.
100 Buxton, *The Redstart*, 2.
101 Ibid., 3.
102 Ibid., 3.
103 Ibid., 4.
104 Nethersole-Thompson, *The Greenshank*, ix.
105 Ibid., 12–13.
106 Letter from Nat Tracy to R. Preston Donaldson, 29 March 1941, RSPB Library 01.05.72(z), Spey Valley, Cairngorm and Rothiemurchus (1940–53).
107 Nethersole-Thompson, *The Snow Bunting*, vii.
108 Nethersole-Thompson, *The Greenshank*, 195.
109 Ibid., 12.
110 Caroline Nethersole-Thompson and Desmond Nethersole-Thompson, 'Greenshank Saga', *Country Life*, 14 December 1940, 760–1.
111 Ibid., 760.
112 Nethersole-Thompson, *The Greenshank*, 44, 72, 64, 151, 155.

113 Nethersole-Thompson and Nethersole-Thompson, 31.
114 Nethersole-Thompson, *The Greenshank*, 58.
115 Ibid., 67–8.
116 Nethersole-Thompson was not unique in handling and stroking wading birds. Edvard K. Barth described touching, stroking, and lifting tame waders on their nests in Scandinavia. See Barth, 'The Tameness of Some Scandinavian Waders'.
117 Nethersole-Thompson, *The Greenshank*, 66–67, 69, plate 2.
118 Nethersole-Thompson, *The Snow Bunting*, 29.
119 Ibid., viii.
120 Nethersole-Thompson, *The Greenshank*, 12.
121 Nethersole-Thompson, *The Snow Bunting*, 6.
122 Nethersole-Thompson, *The Greenshank*, 83; Nethersole-Thompson and Nethersole-Thompson, *Waders*, 23.
123 Nethersole-Thompson, *In Search of Breeding Birds*, 5.
124 Letter from Nat Tracy to R. Preston Donaldson, 9 March 1940, RSPB Library 01.05.72(z), Speyside Watchers' Reports.
125 Letter from N. Tracy to R. Preston Donaldson, 9 March 1940, RSPB Library 01.05.72(z), Speyside Watchers' Reports.
126 Letter from Julian Huxley to Donaldson, 15 March 1940, RSPB Library 01.05.72(z), Speyside Watchers' Reports.
127 Geoffrey Dent letter to Donaldson, 25 March 1940, RSPB Library 01.05.72(z), Speyside Watchers' Reports.
128 Carrie Nethersole-Thompson to Donaldson, 26 September 1945, RSPB Library 01.05.72(z), Speyside Watchers' Reports.
129 Desmond Nethersole-Thompson, 'Preliminary Report on Spey Valley Watching in 1953', RSPB Library 01.05.72(z), Speyside Watchers' Reports.
130 Nethersole-Thompson, *Highland Birds*, 29.
131 George Waterston, 'Report No. 1, 22 April 1957, Ospreys in Speyside', RSPB Library 01.05.709.
132 Nicholson, 'The Trust – Origins and Early Days', 9, 16.
133 Harrisson and Hollom, 'The Great Crested Grebe Enquiry, Part 1', 64.
134 Mandler and Pederson, *After the Victorians;* Mort, *Capital Affairs*.
135 Nicholson, 'The Trust – Origins and Early Days', 28.
136 Cited in Nicholson, 'The Art of Bird Watching', *The Auk* XLIX (1) (January 1932): 119.
137 On Ratcliffe's early enthusiasm for nest finding and egg collecting see Ratcliffe, *In Search of Nature*. Desmond Nethersole-Thompson's son, Des, offered a defence of the 'nest hunters' in chapter 11 of

the volume celebrating the life and work of Derek Ratcliffe: see Nethersole-Thompson, 'Contributions to Field Ornithology'.

138 D.A. Ratcliffe and I.J. Ferguson-Lees, 'Draft Letter Asking for Assistance with the Census of Breeding Peregrines in 1961', 1960, BTO Peregrine Inquiry 1960–1, BTO Archives B1 BTO, Folder 1; Letter from D.A. Ratciffe to Humphrey ap Evans (Falconer), 15 August 1960, BTO Peregrine Inquiry 1960–1, B1 BTO, Archives Folder 1.

139 Burrow, *Passion for Birds,* 106.

CHAPTER TWO

1 'Bird Watching for Juniors', RSPB Film Unit, 1954, RSPB Digital Library.

2 Formed in 1889, the Society had 20, 000 members by 1898, with 152 branches including one in the United States and one in Germany. The Importation of Plumage (Prohibition) Act was passed in 1921 following a campaign led by the Society: see Bassett, 'A List of Historical Records of the Royal Society for the Protection of Birds', National Archives NRA 24, 476, ii–iv; Samstag, *For. the Love of Birds,* 16–53.

3 RSPB Membership Milestones, RSPB Library 01.04.01; Samstag, *For the Love of Birds,* 149; RSPB, 'Facts about the RSPB' (Sandy: RSPB 1977), 11.

4 'The Countryside Under Invasion', *Country Life,* 21 April 1966, 930–2. See Board of Trade, *Staggered Holidays, 1962–3,* Cmd 2105; *Leisure in the Countryside, 1965–6,* Cmd 2928; Countryside Commission, *The Planning of the Coastline: A Report on a Study of Coastal Preservation and Development in England and Wales* (London, 1970); Countryside Commission, *Coastal Recreation and Holidays* (London, 1969); Curry *Countryside Recreation,* 19; Cullingworth, 'Planning for Leisure', 15–16; Dower, *The Fourth Wave.* On bird feeding and its growth, see Jones, *Birds at My Table,* 54–6; Plummer, Risely, Toms, and Siriwardena, 'The Composition of British Bird Communities is Associated with Long-term Garden Bird Feeding'.

5 Lowe and Goyder, *Environmental Groups in Politics,* 10; W.J. Southcombe Ltd, 'RSPB, Research among Members', September 1968, RSPB Library 01.02.70; *Birds* Readers TGI Survey 1993, RSPB Library 01.02.70.

6 BTO, *Annual Reports,* 1944, 7; 1951, 1, 3; 1971, 5; 1976, 5.

7 Spencer, 'The Trust from 1951 to 1982', Hickling (ed.), *Enjoying Ornithology,* 31.

8 The Society's growth as an organization also outstripped that of the county-based birdwatching and naturalists trusts that expanded in the post-war decades. By the late 1970s, the latter groups had a combined membership of 40, 000, with the average county bird society having around 500 members and with a crossover between their memberships and that of the RSPB: see 'County and Regional Bird Clubs and Societies', *British Birds* 70 (12) (1977): 515–21. The Scottish Ornithological Club had been formed in 1937 with sixty-four members. By 1978, it had nearly 3, 000: see letter from George Waterston to E.M. Nicholson, 22 May 1978, E.M. Nicholson BTO Archives R4. Essex Bird Watching and Preservation Society was formed in 1949 and by 1962 had 596 members: see *Essex Bird Report*, 1962. Kent Ornithological Society was formed in 1951 and had 400 members by 1961, *Kent Bird Report*, 10, 1961.

9 'Suggested Amendments to the Protection of Birds Act, 1954', Advisory Committee on the Protection of Birds for England and Wales and Nature Conservancy for Scotland, Nature Conservancy Meeting, 3 November 1960, 1, National Archives (NA), CAB 124/2628; 'The New Protection of Birds Act', *British Birds* XLVII (December 1954): editorial, 409.

10 See 'Tawny Pipit', directed by Bernard Miles and Charles Saunders, 1944; Allsop, *Adventure Lit Their Star*, Davis, 'Britain an Island Again', chapter 4.

11 Lord Tweedsmuir, House of Lords, 29 April 1954, Hansard, 187, page col. 201–72.

12 Sheail, *Nature in Trust*, 22–36.

13 Ibid., 35.

14 Lady Tweedsmuir, Commons Sitting 4 December 1953, Hansard 521, 1,483–584.

15 Ibid.

16 Ibid.

17 Tufton Beamish, Commons Sitting 4 December 1953, Hansard 521, 1,483–584.

18 Sir Hugh Lucas-Tooth, Commons Sitting 4 December 1953, Hansard 521, 1,483–584.

19 Earl of Ilchester, Lords Sitting 29 April 1954, Hansard 187, 201–72.

20 Lord Hurcombe, Lords Sitting 29 April 1954, Hansard 187, 201–72.

21 Earl Jowitt, Lords Sitting 29 April 1954, Hansard 187, 201–72.

22 Arthur Moss, Letter to the Editor, *The Times*, 21 March 1955, 11.

23 RSPB, 'Schoolboy Birds' Nesting', 1955, RSPB Library 04.05.00

24 Letter from Peter Conder to Miss Thomas, 12 February 1957, RSPB Library 04.05.00.

25 'Songbirds' Eggs Protected', *The Times*, 24 April 1963, NA, CAB 124/2628.

26 P.E. Brown and E.R. Parrinder, 'Report of the "Dick Barton" Sub-committee', n.d.; Selected correspondence, 'Breaches of the 1954 Act', 1954–58, 04-05-00; 04-01-00; 04-05-01.

27 Advert for F. Whetmough, Naturalist, Cheshire, Season 1954; Letter from C.H. Gowland, Naturalist; Price list from Watkins and Domester, RSPB Library, 04-05-01; 04-05-00.

28 'Birds Eggs Taken Possession by the Police', Nottingham City Police, Southern Division, 14 May 1957, Hoard of Eggs, RSPB Library 04-05-01.

29 These ranged from £5 to £25 depending on the species.

30 Porter, 'Law Enforcement Five Year Development Plan', 14 September 1971, RSPB Council 3/71/74, RSPB Library 04-01-30.

31 Ibid.

32 Geoffrey Grigson, 'A Matter of Eggs', *Country Life*, 23 August 1956, 388–9. See also 'A Countrywoman's Notes', 8 November 1952, 1475; 'A Countryman's Notes', 16 January 1953, 141; 'Unprotected Eggs', 17 March 1955, 724.

33 'Nests and Eggs of British Birds', *Country Life*, 24 February 1955, 580. 'Shell Nature Study 21: Birds' Eggs', *Country Life*, 13 September 1956, 537.

34 On the continuing problem in the late 1960s and 1970s, see 'Comment: Young Egg Collectors', *Birds* (May/June 1969): 212; 'The Gentle Persuaders', *Birds* (September/October 1975): 9.

35 Mass Observation Bird Nesting Directive 1951, MO 30-1-A-E. The schoolchildren were aged between seven and fourteen.

36 'The Bird Nesters', draft, n.d., n.p., MO 30-1-C; MO 30-1-A-E. On the social profile of mass observers, see Hinton, *Mass Observers*, 268–72.

37 'The Bird Nesters', draft, n.d., n.p., MO 30-1-C; MO 30-1-A-E; MO 96 Panel Number 1096, MO 30-1-A-E. This analysis is based on a sample of the first seventy adult panelists and the summary report produced by MO.

38 MO 96 Panel Number 1096, MO 30-1-A-E.

39 MO 96 Panel Number R4638, MO 30-1-A-E.

40 MO 96 Panel Number 665, MO 30-1-A-E.

41 MO 96 Panel Number 1788, Male Aged 43, MO 30-1-A-E; MO 96 Panel Number N.5, MO 30-1-A-E.

42 MO 96 Panel Number 1831.
43 MO 96, Panel Number 595; MO 96 Panel Number blank.
44 MO 96 Panel Number 054; Panel Number 047; Panel Number R854; Panel Number 1867; Panel Number 624; Panel Number 1899; Panel Number 0222; Panel Number 4495; Panel Number 4633.
45 MO 96 Panel Number 2074.
46 'The Bird Nesters', Draft Report, 1–5, incomplete, n.d, n.p.
47 Ibid.
48 *Junior Bird Watcher*, 1 (2) (December 1955): 2.
49 *Junior Bird Watcher* 3 (1) (July 1960): 3. The JBRC became the Young Ornithologists Club in 1965. By 1980 it had 100, 000 members: see Samstag, *For the Love of Birds*, 149.
50 Preface, P.E. Brown, *Junior Bird Watcher* 3 (1) (July 1960): 3–4; K.G. Spencer, 'A Few Ideas', *Junior Bird Watcher* 4 (1) (February 1962): 2–4.
51 Brown, *Junior Bird Watcher* 3 (1) (July 1960): 4.
52 Hartley, *Bird Watching and Bird Recording* (Sandy: RSPB Occasional Paper Series No. 29), n.d, 3.
53 Hartley, *Bird Watching*, 3; Hartley, 'Bird Diary 1954, Badingham Rectory, Woodbridge, Suffolk', Box 2–7, Papers of Peter Harold Trahair Hartley (1909–1994), Alexander Library Archive and Manuscript Collection, Bodleian Libraries, University of Oxford.
54 Conder, *Bird Notes* 30 (2) (1962): 58.
55 Editorial, *Bird Notes* 31 (5) (1964): 150; Bartlett, 'Bird Gardening in the Suburbs', *Bird Notes* 31 (5) 1964: 156. This was also a theme in Tony Soper's *The Bird Table Book*, which enjoined its readers to turn their 'bird garden' into 'a thriving nature reserve'. On feeding garden birds in an international context, see Jones, *Birds at My Table*, 22–6, 55–6.
56 'Projects for Members', *Bird Notes* 31(11) (1965): 365.
57 Swoop advert, *Bird Notes* 30 (2) (1962): x. Swoop was launched in 1958. There was a longer tradition of feeding birds kitchen scraps, and Daryl Jones quotes *Punch* from 1910 describing wild bird feeding as a 'national pastime' in Britain: see Jones, *The Birds at My Table*, 55. A number of MO panel members in the 1930s described feeding wild birds in their garden: see 'Day Survey Respondent 426', 12 October 1937 and 17 April 1938; 'Daily Survey Respondent 80', 12 February 1937; 'Daily Survey Respondent 630', 12 January 1938.
58 'Readership Survey', *British Birds* 8 (5) (1988): 204.
59 'A Waterbird's World', *Birds* (September/October 1968): 122–5.
60 RSPB Film Unit, Five Year Plan, 1973, RSPB Library 01-03-30.

61 'Highland Birds', Souvenir Programme, 1958, RSPB Film Unit, RSPB Library 01-03-30.

62 Chris Mylne, 'Memorandum on the Production and use of Cine-Films by the RSPB and Proposals for the Administration of the Film Unit', March 1957, RSPB Library 01-03-30.

63 Ibid.

64 RSPB Council, Film Officers Paper, Five Year Plan, Film Department, 1973, RSPB Library 01-03-30.

65 Royal Festival Hall, RSPB Premiere, *The Times*, 13 February 1988, 17.

66 RSPB Film Sub-committee, 28 June 1957, RSPB Library 01-03-30.

67 Ibid.

68 See, for example, 'Reed Warblers', 1959; 'Swallows', 1963; 'A Waterbird's World', 1963; 'Sea Swallows', 1968, RSPB Film Programmes, RSPB Library 01-03-30.

69 See, for example, 'Return of the Osprey', 1962; 'A Waterbird's World', 1963; 'Osprey', 1979, RSPB Film Programmes, RSPB Library 01-03-30.

70 'High Life of the Rook', 1972, RSPB Film Programmes, RSPB Library 01-03-30.

71 See, notably, 'Torrey Canyon', 1968; 'Osprey', 1979; 'Wilderness Is Not a Place', 1973, RSPB Film Programmes, RSPB Library 01-03-30.

72 See, inter alia, 'A Welcome in the Mud', 1973, 'Reserved for Birds', 1961; 'Where the Curlew Calls', 1969, 'Birds of the Grey Wind', 1970; 'The Lonely Level', 1971, RSPB Film Programmes, RSPB Library 01-03-30.

73 Gouyon, *BBC Wildlife Documentaries*, 20; Letter from Philip Brown (RSPB) to Mr George, 4 May 1950, BBC Written Archives, R46/364/2.

74 Parsons, *True to Nature*, 90–4, 367–8.

75 BBC Natural History Unit, Report by Head of West Regional Programmes (Desmond Hawkins), 1962, 2, BBC Written Archives, T31/385, TV Staff 1957–64.

76 Fisher, 'Foreword'; Sielmann, *My Year with the Woodpeckers*, 10.

77 Ibid.

78 Parsons, *True to Nature*, 34; Gouyon, *BBC Wildlife Documentaries*, 27.

79 Look's winter series in 1961 averaged an RI of eighty, making it among the most appreciated programmes produced by the BBC: see BBC Natural History Unit, 1962, BBC Written Archives, T31/385.

80 Huw Welden, Assistant Head of Talks, Television, BBC Natural History Unit, 11 October 1962, BBC Written Archives, T31/386. On the BBC's genres of wildlife television, see Gouyon, *BBC Wildlife Documentaries*; Davies, 'Networks of Nature'.

81 Desmond Hawkins, BBC Natural History Unit, 1962, BBC Written Archives, T31/385.

82 On the BBC's bird-centred documentaries, see Davies, 'Networks of Nature', Appendix C, 237–51.

83 Audience Research T16/40/300/304/306; *Life on Earth*, General WE17/54/1, BBC Written Archives.

84 Davies, 'Networks of Nature', Appendix C, 237–51.

85 Conder, 'Bird Watching as Recreation', *Birds* (July/August 1970): 78.

86 'Christmas Bookshop', *British Birds* 70 (11) (November 1977): 489–93.

87 *The Book of British Birds* (London: Drive Publications, 1969).

88 *Birds* (April 1970): 48.

89 *Book of British Birds*, 15.

90 Book Reviews, *Birds* (November/December 1966): 127.

91 Conder, Review, *Birds* (March/April 1970): 48.

92 *Birds* (January/February 1971): 149.

93 Tory Peterson quoted in the review in *The Auk*: 'A Field Guide to the Birds Giving Field Marks of All Species Found in the Eastern North America', *The Auk* L (1) (1933): 407.

94 Kastner, *A World of Watchers*, 200.

95 On Tory Peterson and the field guides, see Rosenthal, *Birdwatcher*, chapters 2–4.

96 Huxley, 'Introduction', Tory Peterson, Mountfort, and Hollom, *A Field Guide to the Birds of Britain and Europe*, v–vi.

97 Advert, *Birds* (July/August 1966), iii.

98 Tory Peterson, Mountfort, and Hollom, *A Field Guide*, x, xix.

99 Ibid., xix.

100 Ibid., Preface, x; J. Law and M. Lynch, 'Lists, Field Guides, and the Descriptive Organization of Seeing', 282. In a review in the spring 1954 edition of *Bird Notes*, Philip Brown, the head of the RSPB's Watchers Committee, praised the book, describing it as 'as indispensable to the bird watcher as a pair of field glasses': see *Bird Notes*, 26 (2) (1954): 47.

101 Peter Conder, 'Lecture 1: Bird Watching', 8 February 1974, Conder II Box 8: 45–66, Papers of Peter John Conder (1919–1993), Alexander Library and Manuscript Collection, Bodleian Libraries, University of Oxford.

102 Conder, *RSPB Guide to Birdwatching*, 10.

103 Ibid., 60.

104 Ibid., 10.

105 Ibid., 12.

106 Conder's life history of the wheatear conducted on Skokholm formed the basis of his *New Naturalist* monograph.

107 BTO, *Annual Report,* 1960, 11; Conder, 'The Sanders Report and the Naturalists', *Bird Notes* 30 (2) (1962): 56–8. See also 'Present Day Standards of Bird Watching', 19 October 1973; 'Lecture 1: Bird watching', 8 February 1974, Papers of Peter John Conder (1919–1993), Alexander Library and Manuscript Collection, Bodleian Libraries, University of Oxford; Conder, *RSPB Guide,* 63–7.

108 BTO, *Annual Report,* 1949, 43; *Annual Report,* 1951, 33.

109 W.J. Southcombe Ltd, 'RSPB, Research Among Members', September 1968, RSPB Library 01.02.70.

110 Lack, 'The Relationship of the Edward Grey Institute for Amateur Ornithology and the Future of the BTO: A Note for Discussion', 1948, E.M. Nicholson BTO Archives Correspondence Lack, Tinbergen R4.

111 Conder, *Bird Notes* 30 (2) (1962): 58.

112 David Lea, Review, *Birds* (November/December 1967): 265.

113 Ibid., 265; David Lea, 'A Bird in the Bush', *Birds* (March/April 1968): 181. The review of Gooders in *British Birds* made a similar point, conceding that 'some bird watchers may wish that this book should never have been written', though it was more conciliatory suggesting that 'responsible bird watchers [must discipline] their freedoms in the light of the needs of their quarry': see *British Birds* 61 (1) (January 1968): 89–90.

114 Conder, *RSPB Guide,* 10–11.

115 Conder, 'Present Day Standards of Bird Watching', October 1973, 11, Papers of Peter John Conder (1919–1993), Alexander Library and Manuscript Collection, Bodleian Libraries, University of Oxford.

116 *Birds* (January/February 1971): 79. See also 'A Destructive Minority', *Birds* (July/August 1972): 101–2; 'A Disturbing Minority', Letters, *Birds* (July/August 1973): 282.

117 *British Birds* 53 (1) (1960): 3.

118 *British Birds* 53 (4) (1960): 153.

119 *British Birds* 62 (1) (1969): 2–3.

120 Wallace, 'The First 10 Years of the Rarities Committee', *British Birds* 63 (3) (1970): 113–129.

121 Wallace, 'An October to Remember on St Agnes in 1971', *British Birds* 65 (5) (1972): 208–21. For another insider account of rarity chasing, see Oddie, *Bill Oddie's Little Black Bird Book.*

122 Quoted in Rosenthal, *Birdwatcher,* 34.

123 W.J. Southcombe Ltd, 'RSPB, Research Among Members', September 1968, RSPB Library 01.02.70.

124 Conder, Comment, *Birds* (March/April 1972): 24.

125 Tanselle, 'A Rationale for Collecting', 7.

126 Ibid., 8.

127 Kenneth Williamson, 'Finding Out about the Common Birds', *Birds* (January/February 1968): 15.

CHAPTER THREE

1 Council for Nature, *The Observer Wildlife Exhibition*, 1963, CFN 9/3/1, Council for Nature Archives, Linnaean Soiety (Council for Nature archives); Sheail, *Nature Conservation in Britain*, 129.

2 *Fitter's Rural Rides: The Observer Map-Guide to the Countryside*, CFN 9/3/1, Council for Nature Archives.

3 Council for Nature, *The Observer Wildlife Exhibition*, 1963, 26, CFN 9/3/1; Sheail, *Nature Conservation in Britain*, 129–30; Matless, Watkins, and Merchant, 'Nature Trails', 97–131.

4 Texts for labels for NNW, RSPB National Nature Week Exhibition 1963; Letter from J. Clegg to Mr L. Lake, PR Division, J. Walter Thompson, 9 April 1963, RSPB Library 01.09.09.

5 Axell, *Of Birds and Men*, 139; East Suffolk Local Committee, Minutes of Meeting, 10 February 1952, RSPB Library, 01.22.16.

6 Texts for labels for NNW, RSPB National Nature Week Exhibition 1963, RSPB Library 01.0.09.

7 See Mabey, *The Common Ground*, 123–5, 162–79. Mabey's analysis tends to overlook the continuities in bird protection practices.

8 Conder, 'The Osprey's Nest', Letter to the Editor, *The Times*, 24 April 1971, 17.

9 Matless, *Landscape and Englishess*, 254

10 Ibid., 259.

11 Matless, Watkins, and Merchant, 'Nature Trails', 98.

12 'Romney Marsh Bird Sanctuary', *Bird Notes and News* (Spring Number 1930): 15–18.

13 Ibid., 16.

14 Watchers' Committee Reports, 1911–39; Watchers' Bulletins, Nos 1–5, 1954–55, RSPB Archives, 01.05.01.

15 'Romney Marsh Sanctuary', 16.

16 See, inter alia, 'The Protection of Birds', *Bird Notes* (Winter 1955/6): 2–4; Editorial, *Bird Notes* 31 (1) (1964): 3; 'Bird Protection and

Conservation', *Bird Notes* 31(1) (1964): 10; Editorial, *Bird Notes* 31 (2): 42; Editorial, *Birds* (March/April 1966): 22; 'Balranald – A New Bird Reserve in the Outer Hebrides', *Birds* (July/August 1966): 71.

17 Axell, *Minsmere*, 7.

18 Sheail, *Nature Conservation in Britain*, 6.

19 Letter from G. Dent to R. Preston Donaldson, 25 March 1940, Spey Valley, RSPB Library, 01.13.01.

20 Ibid.

21 Sheail, *An Environmental History*, 127

22 Sheail, *Nature Conservation in Britain*, 19; Memorandum No. 3, *Nature Conservation in Great Britain*, Conference Report of the Nature Reserves Investigation Committee, March 1943, 3–5, RSPB Library 03.01.32.

23 Bocking, 'Conserving Nature and Building a Science'; Tansley, *Our Heritage of Wild Nature*.

24 As Dent proudly asserted, the government had accepted NRIC's recommendations 'in toto'. Dent, 'The Problem of Protection', *Bird Notes* 3 (Summer 1950): 81.

25 Sheail, *An Environmental History*, 129. Nicholson also encouraged the voluntary conservation sector to assist in establishing nature reserves. By 1970 the number of nature reserves owned by the county naturalist trusts, and supported by SPNR, had grown to 400. There had been just thirty-four in 1958: see Sussex Wildlife Trust, '100 years of The Wildlife Trusts: A Potted History', 2012, 3.

26 Nicholson, 'Advances in British Nature Conservation', in *Nature Conservancy Handbook*, 1965, 25.

27 Nicholson, 'Advances in British Nature Conservation', 25.

28 NRIC Report No. 2, October 1942, 2, RSPB Library 03.01.32; Sheail, *An Environmental History*, 129.

29 Axell, *Minsmere*, 7.

30 'Report on Areas Suitable for Nature Reserves on the Suffolk Coast, Memo to Minister of Town and Country Planning', 20 October 1943; 'NRIC Suggested Nature Reserve Form, Scott's Hall, Minsmere, Suffolk', handwritten form completed by G. Dent, October 1943, 03.01.32; 'Minsmere Level NR54, Notes on a Meeting Held on 29 November 1947 at the Dolphin Hotel, Thorpeness', RSPB Library 01.05.20 (M2). (Dent was present along with Max Nicholson and Arthur Tansley.)

31 Davies, 'The Return of the Avocet', 164.

32 Ibid., 165.

33 Stanford, *A Bewilderment of Birds* 1954; Davis discusses Stanford in some detail in 'Britain an Island Again', chapters 4 and 5, and in Davis, 'Militarised Natural History', 276–332.

34 Stanford, *A Bewilderment of Birds*, 192–4, 198–208.

35 Ibid., 208.

36 Axell, *Of Birds and Men*, 1992, 59–74.

37 Garlick, *Osprey Involvements*, 81

38 Davis, 'Britain an Island Again', 128.

39 Or perhaps, as a vociferous critic opined, because of the provocation of the RSPB's surveillance: see letter from Eric Hardy to A. Macmillan, editor of *Scottish Birds*, 11 December 1965. Senior figures within the RSPB, including Peter Conder and George Waterston, saw Hardy, who was secretary of the Merseyside Naturalists Association and wrote articles on wild birds for the sporting press, as 'more dangerous than egg collectors': see letter from G. Waterston to P. Conder, 7 October 1965, RSPB Investigations, RSPB Library 01.08.01-6.

40 Letter from P. Brown to Capt. Ogilvie, 17 July 1952, RSPB Library 01.22.16.

41 G. Dent, Watchers' Committee, 'The Management of RSPB Reserves', 1956, RSPB Library 01.05.20 (M2).

42 Minsmere Nature Reserve, Warden's Report on the Breeding Season, 1959, 4, RSPB Minsmere Archives.

43 Minsmere Nature Reserve, Warden's Report on the Breeding Season, 1959, 4.

44 A close-up photographic portrait of an avocet sitting on a nest on the front cover of the autumn 1950 issue of RSPB members' magazine, *Bird Notes*, boldly proclaimed the location of the avocets on Havergate Island. The BBC broadcast a twenty-minute programme on the BBC Home Service called 'The Return of the Avocet' over the Christmas period in 1950. *Bird Notes* (Autumn 1950): 192. *The Times* ran a photo spread on the birds: 'An Avocet Colony in Suffolk', *The Times*, 13 September 1950, 10.

45 It was a notable public relations coup. The national press covered the news and the BBC broadcast a programme, 'The Return of the Osprey', on Sunday 23 August, 1959, Parsons, *True to Nature*, 94. The decision to make the nest site public and invite sightseers to view the birds was not uncontroversial within the RSPB. Some within the Watchers' Committee worried that the publicity would bring along 'rubber-neckers, photographers and impetuous journalists': see 'Operation Osprey', 1959, RSPB Library, 01.05.709. While the RSPB was

happy to orchestrate public encounters with Loch Garten's ospreys, it kept tight control over information about other breeding pairs in Scotland. See the letter from George Waterston to Philip Brown, 20 April 1963, and the reply, RSPB Library, 01.05.709. It was not until 1969 that news of a second pair of nesting ospreys was made public – not by the RSPB, however, but by the Scottish Wildlife Trust at Loch of Lowes. See Garlick, *Osprey Invovlements*, for a detailed account of the RSPB's protection and exhibiting of the Speyside ospreys.

46 Axell, *Minsmere*, 42

47 Minsmere Nature Reserve, Warden's Report on the Breeding Season, 1959, 5.

48 Letter from H. Axell to Peter Conder, 14 June 1960, RSPB Library 01.05.20 (M2).

49 Minsmere Nature Reserve, Warden's Report on the Breeding Season, 1959, 5; RSPB Minsmere Bird Reserve Warden's Report for 1961, 1; 'Beach and Dunes', Minsmere Bird Reserve, Warden's Report for 1963, 7.

50 Letter from Mr Watt to Peter Conder, 20 March 1961, 01.05.20 (M2).

51 Letter from Peter Conder to Westleton Council, 26 March 1961, 01.05.20 (M2).

52 Minsmere Bird Reserve, Warden's Report for 1963, 7.

53 Minsmere Bird Reserve, Warden's Report for 1967, 5; Warden's Report, 1962, 10; Warden's Report, 1973, 30.

54 Minsmere Bird Reserve, Warden's Report, 1972, 5.

55 Minsmere Bird Reserve, Warden's Report for 1967, 5.

56 Axell, *Of Men and Birds*, 125.

57 Minsmere Bird Reserve, Warden's Report for 1973, 6.

58 Minsmere Nature Reserve, Warden's Report on the Breeding Season, 1959, 4; Axell, *Minsmere*, 222.

59 Axell, *Of Birds and Men*, 138.

60 Letter from P.L. Garrett to Philip Brown, 31 May 1960. See the letter from Guy Mountfort asking for a special tour for the wife of the president of the Du Pont corporation, Crawford Greenewalt. As Mountfort put it, 'He is an old friend of mine… I will see that the warden is given a substantial recompense'. Letter from G. Mountfort to Philip Brown, 20 December 1961, and reply, 24 February 1961, RSPB Library 01.05.20 (M2).

61 Letter from H. Axell to P. Conder, 10 September 1959. Axell became friends with the actress Joyce Grenfell and her husband, and Grenfell brought some of the 'Aldeburgh set', including Benjamin Britten,

to the reserve. Grenfell was an RSPB member. Axell also hosted members of the Royal Family on a number of occasions, including the Duke of Edinburgh, who piloted a helicopter onto the reserve. Minsmere Warden's Report for 1961.

62 Warden's Report for 1972, 10.

63 Reserves Department, Five-Year Plan, 7 September 1971; Reserves Visiting, Reserves Department, 1977, RSPB Library, 01.05.01.

64 Warden's Report for 1974, 32–3.

65 RSPB Minsmere Reserve, 1981, 3.

66 Minsmere Nature Reserve Annual Report 1977, 4.

67 Axell, *Minsmere*, 42.

68 Warden's Report for 1974.

69 Axell, *Minsmere*, 42

70 Matless, *In the Nature of Landscape*, 109.

71 Axell, 'Minsmere (Part II), Establishment and Management of an Artificial Brackish Lake with Resting Island ("The Scrape")', *Manual of Wetland Management*, 1973, 11.

72 Letter from J. Clegg to Mr L. Lake, PR Division, J.Walter Thompson, 9 April 1963, RSPB Library 01.09.09.

73 Letter from G. Alkinson-Willes to David Lea, 26 November 1965; Letter from Peter Conder to Captain Ogilvie, 1 August 1958; Letter from Peter Conder to C. Cadbury, 11 August 1958, RSPB Library 01.05.20 (M2).

74 Warden's Report, 1964, 1970, 1971, 1972, 1973.

75 Axell, *Of Birds and Men*, 151.

76 Axell, 'Minsmere (Part II)', 11.

77 P. Brown, *Bird Notes* (Autumn 1948): 155–7.

78 Havergate log book, 1951, RSPB Library, 01.05.16.

79 *Bird Notes* (Autumn 1950): 189.

80 Hosking, *An Eye for a Bird*, 149.

81 Ibid., 142. See also Thomson, *Birds from the Hide*. Describing his experience of photographing Montagu's harriers from a mobile hide, Thomson said 'I had seen and enjoyed the intimate life story of a species', 27.

82 Hosking and Smith, *Fighting Birds*, 105–6.

83 Hosking and Smith, *Fighting Birds*, plate 31a. (See also plates 30a, 30b, 31b.)

84 Hosking and Smith, *Fighting Birds*, 108. (See also 83, 90.)

85 Havergate log books, April–September 1951; April–December 1952; January–October 1953, RSPB Library, 01.05.16. Watchers at Loch

Garten kept even more assiduous records. For the watching seasons of 1959 and 1960, twelve quarto-sized notebooks each containing approximately 200 pages were filled with records of behaviour, including the number of times coitus took place. G. Waterston, *Bird Notes* 29 (5) (1960–61): 135.

86 Garlick, *Osprey Involvements*, 184.

87 Havergate log book, 24 August 1951, RSPB Library, 01.05.16.

88 *The Times*, 2 June 1953, 3; see also Davis, 'Britain an Island Again', 112.

89 *The Times*, 2 June 1953, 3.

90 Axell, *Of Birds and Men*, 137–8.

91 Huxley noted: 'The effect was charming, and reminded me forcibly of the little run made by Macheath and Polly in the "Beggars Opera"', *British Birds (*1925): 90.

92 Cited by Davis in 'Britain an Island Again', 136.

93 Garlick, *Osprey Involvements*, 69.

94 Axell, *Of Birds and Men*, 153.

95 C. Tubbs, 'Viewpoint', *British Birds* 69 (5) (1976): 164.

96 Warden's Report, 1967, 3.

97 Warden's Report, 1960, 10.

98 Minsmere Nature Reserve, Annual Report 1976, 14.

99 See also Davis, 'Britain an Island Again', chapter 4.

100 *Country Life*, 8 February 1930, 190.

101 *Bird Notes and News* (Spring 1930): 16.

102 'The old country scenes are disappearing… From a variety of creatures living in a picturesque harmony to a monoculture is the drift of the modern age', *A Brief Guide to the 13 Sanctuaries of the RSPB*, 1962, Bird Reserves, RSPB Library, 01.05.01.

103 RSPB Council, 1970; *Avocet Island*, RSPB Films, 1950.

104 RSPB Films Programme, 1973.

105 Axell, *Of Birds and Men*, 165, 176.

CHAPTER FOUR

1 Draft letter asking for assistance with the census of breeding peregrines in 1961, D.A. Ratcliffe and I.J. Ferguson-Less, 1960, BTO Peregrine Enquiry, 1960-1, BTO Archives B1; D.A. Ratcliffe, Peregrine Enquiry Progress Report to 27 March 1961 , BTO Archives B1; D.A. Ratcliffe, Peregrine Enquiry Progress Report to 6 July 1961, BTO Archives B1.

2 Ratcliffe, 'The Status of the Peregrine in Great Britain', *Bird Study*, 65.

3 Ratcliffe, 'The Status of the Peregrine in Great Britain', *Bird Study*, 73.

4 Sheail, 'Pesticides and Nature Conservation', 1985, 23.

5 Carson, *Silent Spring*, 11–12; Cramp and Conder, BTO/RSPB Committee on Toxic Chemicals, Report No. 1.

6 *Bird Notes* (Autumn 1960): 91.

7 *Bird Notes* (Summer 1961): 193.

8 *Bird Notes* (August 1962): 223; Carson, *Silent Spring*, 12.

9 Carson, *Silent Spring*, 87.

10 Springer, 'Insecticides – Boon or Bane?', 128.

11 Springer, 'Insecticides – Boon or Bane? (Part Two)', 177.

12 Concerns had first been expressed as early as 1945 by members of NAAS: see Dunlap, *In the Field Among the Feathered*, 222, n. 11.

13 'Poison', *Audubon Magazine* (July–August 1957): 169; 'DDT Spray Alarms Audubon Society', *Audubon Magazine* (July–August 1957): 169; 'Poisons on the Land', *Audubon Magazine* (March–April 1958): 68–9; de Witt, Fish and Wildlife Service, 'Effects of Chemical Sprays on Wildlife', *Audubon Magazine* (March–April 1958): 71; 'The Greatest Killing Programe of All?', *Audubon Magazine* (November–December 1958): 254–6, 294; George J. Wallace, 'Insecticides and Birds', *Audubon Magazine* (January–February 1959): 10–12, 35; 'A Protest Against Spraying', *Audubon Magazine* (July–August 1959): 153–4; 'Late News from the Fire Ant Front', *Audubon Magazine* (March–April 1960): 54–5; 'Our Work on the Pesticide Problem', *Audubon Magazine* (March–April 1960): 65; 'Is Spraying Necessary to Halt Dutch Elm Disease?', *Audubon Magazine* (July–August 1961): 208.

14 Baker, 'Insecticides are Threat to Humans and Wildlife', *Audubon Magazine* (March–April 1958): 75.

15 Letters, 'The Dangers of Poison on the Land', *Audubon Magazine* (May–June 1958): 98.

16 'A Protest against Spraying', *Audubon Magazine* (July–August 1959): 208; Rachel Carson Papers, Yale Collection of American Literature Beinecke Rare Books and Manuscript Library, Yale University, Series 1 Writings 1921–75, Box 31; Letters, 'Thank you, Mr Etter', M. O'Flyn, *Audubon Magazine* (November–December 1959): 242.

17 Wallace, 'Insecticides and Birds', *Audubon Magazine* (January–February 1959): 10–12, 35.

18 See also Wallace, 'Another Year of Robin Losses on a University Campus', *Audubon Magazine* (March–April 1960): 66–9.

19 'Our Work on the Pesticide Problem', *Audubon Magazine* (March–April 1960): 65.

20 News release, Massachusetts Audubon Society, 1962, Carson Papers Series 1 Writings 1921–75, Box 43.

21 *Field Notes*, February 1959, Carson Papers Writings 1921–75, Box 31.
22 'Records of Birds Poisoned by Spray Used on Trees to Combat Dutch Elm Disease', 1962, Carson Papers Writings 1921–75, Box 43.
23 Letter to the Editor, 'Are We Killing Off Birds?', *Caledonian Record*, 11 July 1958, Carson Papers Writings 1921–75, Box 91.
24 'Robins Die by Scores at Bay Point', *Miami Herald*, 21 February 1963, Carson Papers Writings 1921–75, Box 31.
25 C.L. Broley, 'The Plight of the American Bald Eagle', *Audubon Magazine* (July–August 1958): 162–3.
26 Ibid., 163
27 *New York Times*, 13 September 1958, cited in *Audubon Magazine* (November–December 1958): 275.
28 'Continental Bald Eagle Project', *Audubon Magazine* (January–February 1961): 18.
29 On Hawk Mountain see Edwin Way Teale, 'The Long Valley', *Audubon Magazine* (September–October 1956): 20–3, 238; Robert McConnell Hatch, 'Slaughter in Pennsylvania', *Audubon Magazine* (November–December 1956): 256–7.
30 Similar systematic counts of migrating raptors were also undertaken at Cape May Point in New Jersey, with a sanctuary established there in 1935.
31 Broley, 'The Plight of the American Bald Eagle', 163; Carson, *Silent Spring*, 114.
32 Carson, 'Road to the Hawks', 31.
33 On the place of 'wonder' and 'awe' in Carson's writing see, especially, Lear, 'Rachel Carson's Silent Spring'; Sideris, 'Fact and Fiction, Fear and Wonder'.
34 Lear, 'Rachel Carson's Silent Spring', 25, 32.
35 Lyle, *The Gentle Subversive*, 5; Lear, *Rachel Carson*, 347.
36 Memo to Miss Carson from G.J. Wallace, October 12, 1959, Carson Papers Series 1 Writings 1921–75, Box 43; Copy of statement by Dr George J. Wallace to House Subcommittee on Fisheries and Wildlife, Hearing 11502, 3 May 1960, Carson Papers Series 1 Writings 1921–75, Box 43.
37 Rachel Carson to C. Russell Mason, 9 July 1959, Carson Papers Series 1 Writings 1921–75, Box 43.
38 Letter from C. Russell Mason to Rachel Carson, 16 July 1959, Carson Papers Series 1 Writings 1921–75, Box 43.
39 Letter from Rachel Carson to C. Russell Mason, 22 July 1959, Carson Papers Series 1 Writings 1921–75, Box 43.

40 Letter from Harold S. Peters to Rachel Carson, 7 August 1959, Carson Papers Series 1 Writings 1921–75, Box 43.

41 Letter from Harold S. Peters to Rachel Carson, 13 October 1959, Carson Papers Series 1 Writings 1921–75, Box 43.

42 Lear, *Rachel Carson*, 408.

43 Ibid., 426.

44 CBS News Press Release, 14 March 1963; Lear, *Rachel Carson*, Box 88. On the international publication and reception of *Silent Spring*, see 'Silent Spring: An International Best Seller', www.environmentandsociety.org/exhibitions/rachel-carsons-silent-spring/silent-spring-international-best-seller.

45 Lear, *Rachel Carson*, 446–51.

46 Ibid., 451.

47 Lytle, *The Gentle Subversive*, 216–17.

48 In Britain the recent controversy around the drug thalidomide, in which government regulators and pharmaceutical companies were seen to have been negligent about the dangers of the sleeping pill and its catastrophic effects on foetal development, fuelled public suspicion of corporate interests.

49 Carson, *Silent Spring*, 22.

50 Ibid., 22.

51 Ibid., 46.

52 Ibid., 25, 142.

53 Ibid., 7, 570.

54 Ibid., 121.

55 Kinkele, 'The Ecological Landscapes of Jane Jacobs and Rachel Carson'; Horowitz, *Anxieties of Affluence*.

56 Letter from P.E. Howard Jr to Rachel Carson, 19 December 1963, General Correspondence 1942–73, Box 91.

57 Letter from S.L. Showalter to Rachel Carson, 8 August 1963, General Correspondence 1942–73, Box 91.

58 Letter from M.S. Goldman to Rachel Carson, 14 February 1964, General Correspondence 1942–73, Box 91; Letter form A.O. Hage to Rachel Carson, 19 March 1964, General Correspondence 1942–73, Box 91.

59 Letter from H.R. Lillie to Rachel Carson, 18 October 1962, General Correspondence 1942–73, Box 91.

60 Fisher first met Carson in 1953 at Roger Tory Peterson's home. James Fisher, 'The World of Birds', reviewing *Silent Spring*, BBC Home Service, recorded 15 March 1963, transmitted 16 March 1963,

10.10pm–10.45pm, British Reviews, Carson Papers Series 1 Writings 1921–75.

61 Lord Hurcombe, 'The Protection of Nature', *Bird Notes* (Winter 1955/1956): 2.

62 BTO Council Minutes, April 1959, July 1959, December 1959, March 1959; Letter from Robert L. Burnap, NAAS, to David Wilson, BTO secretary, 4 June 1959; Letter from David Wilson to Arnold Nilson, US Fish and Wildlife Service, 11 June 1959, BTO Archives Ken Williamson Papers; W.D. Campbell, 'Purposes of Chemicals Used in Agriculture: A Threat to Wildlife', *Bird Study* 9 (4) (1962): 245–51.

63 'Toxic Chemicals and Birds', *Bird Notes* 29 (3) (Summer 1960): 59; Conder, 'Insecticides and Birds', *Bird Notes* 29 (3): 61–7.

64 BTO Council Minutes, July 1960.

65 Cramp and Conder, *The Deaths of Birds and Mammals*; BTO Council Minutes, April 1961.

66 Ibid., 11

67 Cramp, Conder, and Ash, *Deaths of Birds and Mammals*, 2–7.

68 'Mass Poisoning of Wildlife', *The Times*, 10 April 1962, 13.

69 Cramp, Conder, and Ash, *Deaths of Birds and Mammals*, 2, 13.

70 Ibid., 14.

71 Ibid., 14.

72 Ibid., 8.

73 Ferguson-Lees, 'The Birds of Prey Conference, 1–3 March 1963', *Bird Notes* 30 (7) (Summer 1963): 201.

74 Cramp, 'Toxic Chemicals and Birds of Prey', *British Birds* 56 (1963): 124–37; RSPB Library, 01-06-16.

75 Ibid.

76 BTO, 29th Annual Report, 1962, 1.

77 K. Williamson and R.C. Holmes, 'The Common Bird Census, Methods and Preliminary Results 1960–5', BTO Archives CBC; BTO, Report of the Research Committee to Council, 14 October 1961, BTO Archives CBC.

78 Williamson and Holmes, 'The Common Bird Census', BTO Archives CBC.

79 Ibid.

80 Dobinson and Richards, 'The Effects of the Severe Winter of 1962/3 on Birds in Britain'.

81 S. Cramp and P. Olney, 'The 6th Report of the Joint Committee of BTO–RSPB on Toxic Chemicals with GRA, July 1964–December 1966', 1967, BTO Publication, 10.

82 'To All Members of Scientific Advisory Group re Agenda Item Number 7, Report on the Great Britain Peregrine Census, 1971', 10 December 1970. The census was undertaken by Derek Ratcliffe and the Toxic Chemicals and Wildlife Section at Monks Wood under N.W. Moore, BTO Archives Peregrines B1 F5 1971.

83 R. Spencer, 'The Effects on Birds', *Birds* (July/August 1967): 203; Gill, Booker and Soper, *The Wreck of the Torrey Canyon*, especially chapters 3 and 6.

84 Gill, Booker, and Soper, *The Wreck of the Torrey Canyon*, chapter 6; 'The Poison Sea', A Look Special, No. 159, transmitted 31 March 1967, BBC TV, Peter Scott Papers.

85 'Foulness Objections', *Birds* (July/August 1969): 238; 'The RSPB and Roskill', *Birds* (July/August 1971): 232–3; Letters, *Birds* (January/February 1972): 30. After a Planning Inquiry Commission and much campaigning and deliberation within government, the Labour government abandoned the project in July 1974. See Sheail, *An Environmental History of the Twentieth Century*, 191–6.

86 Philip Larkin's 1972 poem 'Going Going' lamented the loss of the English countryside under the onslaught of tourism, road development, pollution, and urban sprawl. The penultimate stanza reads: 'That will be England gone, the shadows, the meadows, the lanes, the guildhalls, the carved choirs. There'll be books; it will linger on in galleries; but all that remains for us will be concrete and tyres.' Larkin, 'Going Going', cited in Matless, *Landscape and Englishness*, 277–8.

87 *Wilderness Is Not a Place*, RSPB films 1973.

88 Hugh Barrett, *Women's Hour*, BBC Radio 4, recorded 8 June 1967; John Moore to Michael Walter, 19 October 1966; Alan Bold, 'A Demonstration of Summer', *The Times*, 28 June 1969, 22; J.A. Baker Archive, Albert Sloman Library Special Collections, University of Essex, A3/1.2–1.3 General 1966–88. Later commentators have deepened and extended this praise for the book. For a younger generation of nature writers, *The Peregrine* and its enigmatic author have become the touchstone for a new wave of nature writing; Baker, 'The Peregrine', *Audubon Magazine* (September/October 1967): 26–42.

89 Baker, *The Peregrine*, 31–2.

90 Alongside his second book, *The Hill of Summer*, this was the only other piece of published work that Baker produced.

91 Baker, 'On the Essex Coast', 427.

92 Ibid., 428, 431.
93 Letters from M. Muncaster and Mrs M Drake, *Birds* (January/February 1972): 30.
94 Baker, *The Peregrine*, 31–2.
95 Baker, *The Peregrine*, 113.
96 Letters to J.A. Baker from R.S. Morrison, Berkshire, 26 November 1969; David Smith, Worcestershire, 19 March 1974. Letter from Mrs Whitfield Vye, Connecticut, 15 December 1968, J.A. Baker Archive, Albert Sloman Library Special Collections, University of Essex, A6.1-A6.52, fan mail.
97 See also letter from John A.J.C. Verpaalen, Netherlands, 17 July 1970; Letter from Piergiogio Vanossi, Italy, 22 January 1980; Letter from E. Rober-Tissot, Switzerland, 13 January 1970; Letter from Mrs Whitfield Vye, Connecticut, 15 December 1968, J.A. Baker Archive, Albert Sloman Library Special Collections, University of Essex, A6.1-A6.52, fan mail.
98 Letter from C.H. Shafer to Mr John A. Baker, 22 April 1968, J.A. Baker Archive, Albert Sloman Library Special Collections, University of Essex, A6.1-A6.52, fan mail.
99 Letter from John Canright, 19 February 1973; see also letter from David Smith, 19 March 1974, J.A. Baker Archive, Albert Sloman Library Special Collections, University of Essex, A6.1-A6.52, fan mail.
100 Carson, 'Rachel Carson Answers Her Critics', *Audubon Magazine* (September/October 1963): 262–6, 313.
101 A series of voluntary restrictions were introduced in 1964, 1969 and 1976 prior to the full banning: see Newton, *Farming and Birds*, 205.
102 Newton, *Farming and Birds*, 565.
103 *Audubon Magazine* (January/February 1967): 4; Hammond, 'Conservation and Sister Organizations, the RSPB', 162–3.
104 Ferguson-Lees, 'Distribution Maps and Bird Atlases', in Hickling (ed.), *Enjoying Ornithology*, 74–6; Sharrock, *The Atlas of Breeding Birds in Britain and Ireland*, especially 'Foreword' by J. Ferguson-Lees and 'Introduction' by J.T.R. Sharrock, 9–28.
105 Editorial, *Birds* (July/August 1967): 192.

CHAPTER FIVE

1 E.M. Nicholson, 'Nene and Ouse Washes Visit, 7–8 January 1962', NC, WCC, 22 January 1962, NA FT 2/178.

2 On Nicholson's tea parties see Nicholson, 'National Wildfowl Refuges', in Sedgwick, Whitaker and Harrison (eds), *The New Wildfowler in the 1970s*, 217–8. Also cited in Lear, *Man and Wildfowl*, 170; and in Matless, Merchant, and Watkins, 'Animal Landscapes', 194; NC, Wildfowl Conservation Committee, Terms of Reference, Composition and Related Matters, 23 November 1960, NA FT 2/97.

3 Peter Scott, 'Isles of Ely Marshes – Proposed Wildfowl Refuges', Meeting Held at County Hall, 8 January 1962, between Wildfowl Trust, NC, WAGBI, County Planning Officers and Engineers, NA FT 2/97, 3.

4 Earl of Mansfield, 'Foreword', in Sedgwick, Whitaker, and Harrison (eds), *The New Wildfowler*, 5.

5 On the latter tradition see Marchington, *The History of Wildfowling*, chapter 7.

6 On anti-field sports and animal welfare campaigning see McKenzie, 'The Origins of the British Field Sports Society', 177–91; wildfowl were not subject to the Game Laws that protected birds like pheasant and grouse and were deemed to be 'common property'. See Martin, 'Wildfowling and Its Evolution as a Sporting Activity', 119.

7 'Goodbye WAGBI', *Country Life*, 2 July 1981, 4.

8 'A Bill Which Will injure Shooting', *Country Life*, 14 November 1925, 3.

9 In 1957 WAGBI had 174 life members, 200 ordinary members, and 60 affiliated clubs with assets of only £457. This had grown to a membership of 23, 500 by 1972 and swelled again to pass 40, 000 members in 1977: see Marchington, *The History of Wildfowling*, 244, 249. The legislative changes to which WAGBI responded were the Wild Birds Protection Bill (1925), the Wild Birds (Ducks and Geese) Protection Bill (1939), and the Protection of Birds Act (1954). For the organization of WAGBI around the 1925 Bill see 'Unity to Protect Shooting', *Country Life*, 8 August 1925, 38; 'The Wildfowlers Association', *Country Life*, 24 October 1925, 8; 'A Bill Which Will Injure Shooting', *Country Life*, 14 November 1925, 3.

10 The Royal Charter (1904) recorded that the Society would 'take no part in the question of killing game birds and legitimate sport of that character': see Bassett, 'A List of Historical Records of the RSPB', ii.

11 Matless, 'Versions of the Animals – Broadland', 118.

12 'Unity to Protect Shooting', *Country Life*, 8 August 1925, 38; 'The Wildfowlers Association', *Country Life*, 24 October 1925, 8;

13 Innes-Smith, 'James Wentworth Day (1899–1983)'; Wentworth Day, *Sporting Adventure*, author's biography.

14 Wentworth Day, *Sporting Adventure*, 5–6.

15 Martin, 'Wildfowling and Its Evolution as a Sporting Activity', 119–39.

16 Wentworth Day, *The Modern Fowler*, 2–5.

17 Ibid., 8.

18 Ibid., 13.

19 Ibid., 72.

20 Wentworth Day, *Broadland Adventure*, 42–51. On Wentworth Day's moral geography see Matless, *In the Nature of Landscape*, chapter 4.

21 Wentworth Day, *Broadland Adventure*, 42.

22 Ibid., 11.

23 Ibid., 146.

24 Wentworth Day, *Broadland Adventure*, 145–6; *The Modern Fowler*, 31, 70.

25 Wentworth Day, 'Wildfowlers' Moon', *Country Life*, 19 April 1946, 712.

26 Wentworth Day, *Sporting Adventure*, 13.

27 A point also made by J.G. Millais in his introduction to Nichols, *Birds of Marsh and Mere and How to Shoot Them*, 5–7.

28 For a description of punt gunning see Wentworth Day, 'Wildfowlers' Moon', *Country Life*, 19 April 1946, 7,120–212; *The Modern Fowler*, chapter 17.

29 Wentworth Day, 'Big-gunners of the Tideways', *Country Life*, 4 October 1956, 702–4.

30 Wentworth Day, 'Rules for Success in Shore Shooting', *Country Life*, 18 January 1946, 3; *The Modern Fowler*, 189–90.

31 Wentworth Day, 'Rules for Success in Shore Shooting', *Country Life*, 18 January 1946, 3.

32 Wentworth Day, letter to the editor, *The Times*, 9 March 1974, 14; Wentworth Day, *Wild Wings and Some Footsteps*, 200.

33 Nichols, *Birds of Marsh and Mere*, 78.

34 Wentworth Day, *The Modern Fowler*, 32.

35 Wentworth Day, *Wild Wings*, 193.

36 Ibid., 8.

37 Ibid., 8–17, 193–203.

38 Compare, for example, chapters 1, 5, and 6 of *Wild Wings* with chapters 1, 5, and 6 of *British Birds*.

39 Wentworth Day, 'Poison on the Land', *Country Life*, 30 January 1953, 267; Wentworth Day, *The Poison on the Land*.

40 Nichols's 1926 book *Birds of Marsh and Mere and How to Shoot Them* prefigured Wentworth Day in combining the experiences and techniques of wildfowling with elements of the natural history of geese and ducks and identification plates of them that were advanced in their description of birds' field characteristics.
41 Attenborough, 'Introduction', ix.
42 Huxley, *Peter Scott*, ix–xix, 20, 64–7, 84–5.
43 Scott, *The Eye of the Wind*, 170.
44 Ibid., 76.
45 Ibid., 79.
46 Scott, *Morning Flight*, 64.
47 Ibid., 31, 56, 59.
48 Ibid., 47.
49 Scott, *Wild Chorus*, 2; *Morning Flight*, 32.
50 Scott, *The Eye of the Wind*, 94.
51 Scott, *Wild Chorus*, 49; E. Huxley, *Peter Scott*, 90.
52 Scott, *Morning Flight*, 37–8.
53 Ibid., 64.
54 Scott, *The Eye of the Wind*, 78.
55 Ibid., 547.
56 See Rembrandt Harmenszoon van Rijn, 'Hunting Still Life with Girl', 1630; 'Self-Portrait with Bittern', 1639. Still life with dead game and wildfowl was a popular genre within seventeenth-century Dutch painting.
57 See J. Vincent, 'Coot Drive on Lord Desborough's Shoot, Hickling Broad', *Country Life*, 23 December 1935, 190; advert for *The Field*, *The Field*, 28 September 1940, back page; G. Farrar, 'The Ways of Wildfowl – II', *The Field*, 9 November 1940, 602; G. Farrar, 'The Ways of Wildfowl – III', *The Field*, 16 November 1940, 652–3.
58 Frances Pitt, 'Wildfowl: A Review of Recent Work by Peter Scott', *Country Life*, 12 June 1937, 674–5; 'Peter Scott at Home', *Country Life*, 24 December 1938, 621; Frances Pitt, 'The Work of Peter Scott', *Country Life*, 20 November 1935, 563; Frank Wallace, 'The Art of Peter Scott', *Country Life*, 4 July 1936, 8–9.
59 Huxley, *Peter Scott*, 84–5.
60 Scott, *The Eye of the Wind*, 555–6.
61 Ibid., 248–9.
62 Wentworth Day's dog Mr Soapey Sponge was named after the hero of 'Mr Sponge's Sporting Tour', written by the Victorian sportsmen

and novelist Robert Smith Surtees. See Wentworth Day, 'An Historic Marsh – 5', *Shooting Times*, 3 September 1954, 604; Scott, *The Eye of the Wind*, 249.

63 Scott's relationship with 'Annabel' recalled the 'goose children' nurtured by his friend Konrad Lorenz.

64 Scott, *The Eye of the Wind*, 554; The Severn Wildfowl Trust, Report of Council, 1947, 2–4.

65 Scott, *The Eye of the Wind*, 547–8, 555.

66 Marchington, *The History of Wildfowling*, 153.

67 'Preservation of Wild Geese and Wild Ducks', *Nature* 143 (6 May 1939): 754. The recommendations of an international conference held in London between 12 and 14 October 1927 and chaired by Viscount Ullswater fed into the 1939 legislation. See Home Office, 'Report of the International Conference on the Protection of Migratory Wildfowl', HMSO, 1928, RSPB library, 04-01-05.

68 NC, WAGBI, and Wildfowl Trust, *The Story of a Triumvirate*.

69 In 1939 the paper *Shooting Times and British Sportsman* signaled its concern about the obsession with 'bird sanctuaries' from bird protectionists, worrying that the law might soon be used to create wildfowl sanctuaries. 'Let the modern law keep off,' it proclaimed. 'Jottings for Wildfowlers', *Shooting Times and British Sportsman*, 8 July 1939, 841.

70 Scott, Severn Wildfowl Trust, Second Annual Report, 1948–49, 4.

71 Ibid., 3.

72 Ibid., 42; Seventh Report of the Wilfowl Trust, 1953–54, 1. Her Majesty the Queen became the Trust's patron in 1953.

73 Seventh Report of the Wildfowl Trust, 1953–54, 7; Letters to the Editor, Scott, *Shooting Times*, 13 August 1954, 554.

74 Scott, 'The Severn Wildfowl Trust: Its Work and Objects', *Shooting Times*, 26 March 1954, 205–6.

75 Editorial, 'Protection of Birds Bill', *Shooting Times*, 28 November 1953, 757; Letters to the Editor, C. Beresford, 5 December 1953, 782; Editorial, 'Wildfowling Season in Jeopardy', 1 January 1954, 17.

76 'The Wild Birds Protection Bills: Taking Stock', *Shooting Times*, 15 January 1954, 40–1.

77 Tower-Bird, 'Wild Birds in Britain', *Shooting Times*, 12 March 1954, 162.

78 Ibid., 162.

79 Dr King, Third Schedule [of Protection of Birds Act], House of Commons, 9 April 1954, 526 cc 710–44 Hansard; E.L. Parish, Chair of WAGBI, 'The Brent Goose – IV', *Shooting Times*, 26 March 1954, 195.

80 Letters to the Editor, J. Buxton, *Shooting Times*, 2 April 1954, 215. See also hostile letters about Peter Scott and the Severn Wildfowl Trust from James Wentworth Day and Dr G. Storey, Honorary Secretary of WAGBI, Letters to the Editor, 16 April 1954, 246–7.

81 In the amending of the Protection of Birds Act in 1967, wildfowling interests in Parliament were represented by John Farr MP, Marcus Kimball MP, and Sir Tufton Beamish MP, wildfowler and president of the RSPB, Marchington, *The History of Wildfowling*, 200; Parliamentary Debates, House of Commons Official Report, Standing Committee C, Protection of Birds Bill (Lords), 5 April 1967, 302.

82 Editorial, 'In Another Place', *Shooting Times*, 23 April 1954, 257.

83 Letter from Peter Scott to E.M. Nicholson, 28 March 1954, NA, FT 3/491.

84 Letter from E.M. Nicholson to Mr Quick, 14 April 1954, NA, HO 285/162.

85 Third Schedule [of Protection of Birds Act], House of Commons, 9 April 1954, 526 cc 710–44 Hansard.

86 The Protection of Birds Act, 1954, *Shooting Times*, 11 June 1954, 371.

87 Tower-Bird, 'Another Call to Arms', *Shooting Times*, 8 July 1955, 424.

88 Draft Minutes of a Meeting Held at 25, Eccleston Square on 21 October 1957 at 5pm between Representatives of WAGBI and the RSPB, RSPB Library 04-01-55.

89 NC, Scientific Policy Committee, Notes by D-G, EMN, 25 November 1953, NA, FT 3/490.

90 WAGBI Dinner and AGM, *Shooting Times*, 10 June 1965, 727.

91 P. Scott, 'WAGBI's Part in Wildfowl Conservation', *Wildfowler Magazine*, 1960, 29; E.M. Nicholson, 'As Others See Us', *Wildfowler Magazine*, 1961/2, 29–30; P. Brown, 'Conservation – A Must', *Wildfowler Magazine*, 1963/4, 31–2; NC, WAGBI, and Wildfowl Trust, *The Story of a Triumvirate*.

92 Editors, *The New Wildfowler in the 1970s*, 34.

93 NC, WCC, 'The National System of Wildfowl Refuges, An Explanatory Memorandum', I. Prestt, 20 June 1961, NA, NC, FT 2/177; 'Shooting Licenses and Shooting Statistics', H. Boyd, 11 June 1965, NA, NC, FT 2/179; 'Wildfowling at Caerlavarock NNR: The First 7 Years', 30 November 1964, NA, NC, FT 2/179; 'Notes for Wildfowlers Shooting on Lindisfarne NR', NA, NC, FT 2/179; 'Wildfowling on the N. Norfolk Coast', 1 September 1967, NA, NC, FT 2/198.

94 'Conservation and Sport', Editorial, *Shooting Times*, 23 December 1960, 1157; 'A Policy for Sanctuaries', Editorial, *Shooting Times*,

16 September 1960, 839; 'Suicide Tactics', Editorial, *Shooting Times*, 16 December 1960, 1147; 'WAGBI', Editorial, 27 May 1965, 661.

95 P. Brown, 'Bird Protection and the Wildfowler', in *The New Wildfowler*, 238.

96 Ibid., 238.

97 J. Anderton, 'Shooting and Bird Protection', *Bird Notes* 30 (4) (Autumn 1962): 115.

98 P. Conder, 'Comment', *Birds* (March/April 1968): 186.

99 'Breydon Water May Be Bird Sanctuary: Indiscriminate Shooting', *Daily Telegraph*, 8 October 1963, NA CAB 124/2628.

100 Brown, 'Bird Protection and the Wildfowler', 238.

101 'Conservation and Sport', Editorial, *Shooting Times*, 18 February 1965, 213.

102 Anderton, 'Shooting and Bird Protection', 114.

103 P. Conder, 'The RSPB and Roskill', *Birds* (July/August 1971): 232–3.

104 Inter alia, J. Harrison, 'A Day at Cley Marshes', *Shooting Times*, 27 May 1965, 672–3; 'Myvatin, Lake of Midges', *Shooting Times*, 8 July 1965, 864–5; 'Invasions and Eruptions', *Shooting Times*, 18 January 1969, 104–5.

105 Conder, 'Conservation and Bird Protection Part Two', 261–2.

106 J. Harrison, 'The Wildfowler as a Conservationist', *Birds* (March/April 1968): 205.

107 B. Mitchell and B. Hawkes, 'Are Wildfowlers' Conservationists?' *Birds* (September/October 1969): 285.

108 Letters, D.S. Bunn, *Birds* (November/December 1969): 309.

109 Letters, R. Moles, *Birds* (November/December 1969): 309.

110 Letters, S. Cowdry, *Birds* (January/February 1970): 25. See also letters by H. Reid and H. Wilks, Kent Trust for Nature Conservation, 25; Letters, Earl of Mansfield, March/April 1970, 48.

111 Letters, B. Hawkes and B. Mitchell, *Birds* (March/April 1970): 49.

112 Wentworth Day, 'Why Farmers Can No Longer Afford Our Total Protection Laws for Brent Geese', *The Times*, 9 March 1974, 14.

113 Letters to the Editor, P. Conder, P. Scott, and J. Anderton, *The Times*, 13 March 1974, 17.

114 Matless, *In the Nature of Landscape*, 128.

115 'A Visitor Survey at Slimbridge', *Wildfowl* 22 (1971): 126. Around 60 per cent of these visitors used the restaurant facilities provided at the site.

116 C. Tubbs, 'Poor Substitute for Wilderness', *Birds* (Summer 1979), cited in Mabey, *The Common Ground*, 125.

CHAPTER SIX

1 Editorial, *The Falconer* 5 (8) (December 1974): 7; see also, Editorial, *The Falconer* 3(2) (December 1955): 41; M.H. Woodford, 'The 40th Anniversary of the BFC', *The Falconer* 5 (1) (December 1967): 11.

2 Blaine, *Falconry*; Mellor, 'Falconry in Britain', 1.

3 Mellor, 'Falconry in Britain', 1; Campbell and Lack (eds), *A Dictionary of Birds*, 203–6.

4 Beebe and Webster, *North American Falconry*, 297.

5 Wentworth Day, *Sporting Adventure*, 133–4.

6 *The Falconer* 1 (May 1937): 3.

7 Mellor, 'Falconry in Britain', 265.

8 'Falconry Still Popular Sport', *The Times*, 14 November 1957, 7; Robinson, *Falconry in Britain*, 3; Faux, 'Patience of a Falconer', *The Times*, 27 August 1970, 6.

9 Cocker, *Birds and People*, 163.

10 Sylvia Townsend Warner, in her biography of T.H. White, describes how he turned to animals for a 'renewal and enlargement of his being': see Townsend Warner, *T.H. White: A Biography*, 139.

11 'Foreword: Portal of Hungerford', in Woodford, *A Manual of Falconry*, xiii.

12 'Foreword', Blaine, *Falconry*, 8.

13 Blaine, *Falconry*, 14–16.

14 Ibid., 25.

15 Ibid., 87.

16 Blaine mentions a third form, 'gull hawking', *Falconry*, 127–9.

17 In actor James Robertson Justice's account of hawking in Scotland, he describes a number of birds being lost or killed: see J. Robertson Justice, 'Hawking in Scotland', *The Falconer* (3) (December 1958): 155.

18 Blaine, *Falconry*, 161.

19 Blaine, *Falconry*, 229.

20 Woodford, *A Manual of Falconry*, 159.

21 Ibid., 30.

22 Ibid., 77.

23 Ibid., 159–60; Blaine, *Falconry*, 1936, 13.

24 Beebe and Webster, *North American Falconry*, 297.

25 Woodford, *A Manual of Falconry*, 54.

26 I. Neill, 'Goshawks for a Yeomen', *Country Life*, 17 September 1970, 678.

27 J. Stevenson, 'Grouse Hunting in Caithness', *Country Life*, 26 September 1963, 724.
28 Ibid., 726.
29 C. Knight quoted in Fitzgerald, *Falconry's Falconer*, 20.
30 Matless, *Landscape and Englishness*, 166.
31 Wentworth Day, *Sporting Adventure*, 133–6; Bartie, Fleming, Freemen, Hulme, Hutton, and Readman, 'Historical Pageants', 866–902.
32 Blaine, *Falconry*, 237; Evans, *Falconry*, 74–5. Macdonald, *Falcon*, 151–2.
33 Link and Hornburg, 'He Who Owns the Trifels', 208–39.
34 Ibid., 231.
35 *The Falconer* 1 (May 1937): 5, 23.
36 Ibid., 23.
37 *The Falconer* 3 (November 1938): 4.
38 Macdonald, *Falcon*, 99.
39 W. Kennet Richmond, 'Training the Yeoman's Hawk', *Country Life*, 20 September 1956, 586.
40 White had discovered Malory's text while an undergraduate at Cambridge University and wrote his dissertation on it: see Townsend Warner, *T.H. White: A Biography*, 41.
41 White, *England Have My Bones*, 5–6; Montefiore, 'Englands Ancient and Modern', 41.
42 Townsend Warner, *T.H. White: A Biography*, 24.
43 Ibid., 83
44 'Book Reviews', *The Falconer* 2(5) (June 1952): 30. See Lord Tweedsmuir's more supportive review in *The Spectator*: 'Gos and Others', 16 August 1951, 7.
45 Townsend Warner, *T.H. White: A Biography*, 95.
46 White, *The Goshawk*, 15.
47 Ibid., 15.
48 Ibid., 56.
49 Ibid., 134.
50 Ibid., 74–5.
51 Ibid., 75–6, 134.
52 Macdonald, *H Is for Hawk*, 76. Macdonald's book blends an account of her own training of a goshawk with a sustained meditation on White's book and his life.
53 Macdonald, *H Is for Hawk*, 37–8, 118.
54 White, *The Goshawk*, 54; Townsend Warner, *T.H. White: A Biography*, 91.
55 White, *The Goshawk*, 194–5.

56 Townsend Warner, *T.H. White: A Biography*, 129.
57 Ibid., 142.
58 Ibid., 217.
59 Hines, *A Kestrel for a Knave*, 203.
60 The book is dedicated to Hines's brother Richard. Richard Hines kept a trained kestrel called 'Kes' and was an important inspiration for his older brother's story. See R. Hines, *No Way But Gentlenesse*, especially chapters 9–12.
61 This interconnection between the urban streets, industry, and the countryside is captured in Ken Loach and Tony Garnett's cinematic adaptation of the book in their film *Kes* (1969) when Billy is shown reading a comic on a hillside overlooking the pounding industry of a factory.
62 Hines, *A Kestrel for a Knave*, 81.
63 Ibid., 139.
64 Ibid., 146.
65 The phrase is del Valla Alcade's, cited in Dobson, 'Beyond the Dark Satanic Mills', 108.
66 R.B. Treleaven, 'A Peregrine's Eyrie', *The Falconer* 3 (2) (December 1955): 58–61; 'The Non-breeding of Peregrines in the West of England', *The Falconer* 3(5) (December 1958): 158.
67 G. Dent, 'Suggested Eagle and Hawk Adoption Scheme', November 1937, RSPB Library, 04-02-50; *The Falconer* 2 (May 1938): 16.
68 Editorial, *The Falconer* 3 (6) (December 1960): 173.
69 P. Barclay-Smith, 'Birds of Prey Need More Protection', *The Falconer* 3 (6) (December 1960): 188; Editorial, *The Falconer* 4 (1) (December 1961): 5; Editorial, *The Falconer* 4 (2) (December 1962): 41.
70 Letter from Philip Brown to David Wilson, BTO, 29 March 1962, RSPB Library, 04-01-53; M. Woodford, 'Some Reflections on the Conference', RSPB Library, 04-01-53.
71 'Meetings and Conferences: Working Conference on Birds of Prey, Caen, 1964', *The Falconer* 4 (4) (December 1964): 131; Editorial, *The Falconer* 4 (4) (December 1964): 130.
72 *The Falconer* 4 (6) (December 1966): 5.
73 'Report on the IXth Conference of the ICBP, European Continental Section, Hungary, May 1968', *The Falconer* 5 (2) (December 1968): 75.
74 *ICBP Annual Report for 1970*, 25.
75 *ICBP Annual Report for 1970*, 'World Working Group on Birds of Prey, 26–26; British Section, General Report; Report on the 10th Conference of the European Continent Section of ICBP, May 1972.

76 P. Conder, 'Comment', *Birds* (November–December 1970): 124.
77 RSPB, 'Falconry – Should I Take it Up?', Leaflet, Revised March 1979, RSPB Library, 04-01-53.
78 J.G. Mavrogordato, 'The Contribution of the Falconer to the Conservation of Predators', *The Falconer* 5 (6) (December 1973): 47–8.
79 Morley, 'Responsible Falconry', Letter, *Birds* (July–August 1973): 283.
80 J.N.P. Watson, 'Forum for Sporting Ethics', *Country Life*, 22 July 1982, 234–5.
81 S.M. Woodmen, 'Falconry Displays', *Country Life*, 26 October 1978, 1,309.
82 *The Times*, 2 April 1973, 3.
83 Automobile Association, AA *Book of the British Countryside*, 159.
84 'The Peregrine and the Pigeon', RSPB Publication, 1934, RSPB Library 04-01-50.
85 Macdonald, *Falcon*, 114. In the prologue to Michael Powell and Emeric Pressburger's 1944 film *A Canterbury Tale*, a falcon released into the Kentish skies returns as a spitfire as the scene switches from Chaucerian pageant to wartime England.
86 Gruffud, 'Reach for the Sky', 19–24.
87 Ibid., 20.
88 Tory Peterson, *Birds Over America*, 135–48. See the same metaphor and chivalric images of the peregrine in Murphy, *The Peregrine Falcon*, 5, 6.
89 Willcock, 'Foreword', in Brown, *Birds of Prey*, 5.
90 Brown, *Birds of Prey*, 48, 55.
91 RSPB Film Unit, 'Outline and Treatment for Film "Birds of Prey of Europe"', RSPB Film Committee, RSPB Library 010-03-30; 'The Winged Aristocrats', RSPB Film Programme 1970, RSPB Library 010-03-30.
92 RSPB Film Unit, 'Outline and Treatment for Film "Birds of Prey of Europe"', RSPB Film Committee, RSPB Library 010-03-30.
93 Baker, *The Peregrine*, 30.
94 Ibid., 92, 41, 31, 33–4.
95 Ibid., 124.
96 Ibid., 124–5.
97 Ibid., 159.
98 Macdonald, *Falcon*, 8; Macfarlane, 'Introduction', *The Peregrine*, vii, ix, x.
99 Macdonald, *Falcon*, 98.
100 Editorial, *The Falconer* 5 (3) (December 1969): 128.

101 Macdonald, *Falcon*, 275.
102 Robinson, *Falconry in Britain*, 3, 13.

CHAPTER SEVEN

1 'Massacre of the Songbirds', *Sunday Mirror*, 2 October 1969, 1, 12–13.
2 'Massacre of the Song Birds in Italy', *The Guardian*, 2 February 1968, 12.
3 Letter from H.W. Clarke, Staines, Middlesex, to Director, NC, 28 February 1970, NA FT 3/491.
4 *Wild Birds and the Law*, RSPB, 1978.
5 Shrubb, *Feasting, Fowling and Feathers*, 14, 161; 'Song Birds as Food', *The Times*, 1 April 1916, 3; 'Song Birds as Food', *The Times*, 31 March 1916, 9; 'Seabirds Eggs and Human Food', *Country Life*, 22 January 1916, 127.
6 Shrubb, *Feasting, Fowling and Feathers*, 139; Fraser Darling, *Natural History of the Highlands and Islands*, 195.
7 Huxley, 'Birds and Men on St Kilda', 79.
8 Kallsberg, 'The Faeroes Today', in Williamson, *The Atlantic Islands*, 307.
9 Nicholson, 'Preface', in Williamson and Morton Boyd, *St Kilda Summer*, 11.
10 Tom Steel, author of one of the most celebrated paeans to the people of St Kilda and the islands' landscapes, memorably described it as 'a land of superlatives': see Steel, *The Life and Death of St Kilda*, xi.
11 Steel, *The Life and Death of St Kilda*, xi.
12 Huxley, 'Birds and Men on St Kilda', 78.
13 National Records of Scotland, Stories from St Kilda, 'Population', 7.
14 Agriculture and Fisheries Department, file AF57/5, National Records of Scotland, Stories from St Kilda, 'Way of Life', 2; Steel, *The Life and Death of St Kilda*, 56.
15 Steel, *The Life and Death of St Kilda*, 63.
16 Williamson and Morton Boyd, *St Kilda Summer*, 111.
17 Steel, *The Life and Death of St Kilda*, 58; Fisher, *The Fulmar*, 139.
18 Steel, *The Life and Death of St Kilda*, 58–9.
19 See also the (St Kilda) Bill that extended the provisions of the Wild Birds Protection Act (1880), Bill 248, 1904.
20 Fisher, *The Fulmar*, 117, 133, 93.
21 Fisher and Lockley, *Sea-birds*, 82.
22 Williamson and Morton Boyd, *St Kilda Summer*, 31.

23 See chapter 2 in this book, as well as Lockley, 'Bird Migration Stations in Britain', *New Naturalist Journal* (Autumn 1948): 165–71; A. Nicholson (ed.), *Letter from Skokholm,* 23–40; Fisher, 'St Kilda: A Natural Experiment', *New Naturalist Journal* (Summer 1948): 90–108.

24 Huxley, 'Birds and Men on St Kilda', 82.

25 Williamson, *The Atlantic Islands,* 144–5.

26 Fisher and Lockley, *Sea-birds,* 99.

27 Williamson, *The Atlantic Islands,* 149.

28 Ibid., 151.

29 For a fuller account of the hunting of seabirds see Williamson, *The Atlantic Islands,* 146–59; Lockley, *I Know an Island,* 181–193; Norrevang, 'Traditional Seabird Fowling in the Faeroes', 275–81; Baldwin, 'Sea Bird Fowling in Scotland and Faroe', 60–103.

30 Williamson, *The Atlantic Islands,* 146; Fisher and Lockley, *Sea-birds,* 100.

31 Lockley, *I Know an Island,* 187.

32 Ibid., 208, 209, 213.

33 Murray, *The Guga Hunters,* 30.

34 Ibid., 61.

35 MacGeoch, in Bannerman, *Birds of the British Isles,* 8 (1959): 33–4.

36 Macgeoch, 34; Beatty, *Sula,* 98.

37 Sefton Gordon, 'Gannets as food', Letter to the Editor, *The Times,* 30 August 1938, 6; Sefton Gordon, 'Gannet Harvest: A Trip to Sula Sgeir', *The Times,* 15 December 1938, 19.

38 Beatty, *Sula,* 98; Murray, *The Guga Hunters,* 153.

39 Murray, *The Guga Hunters,* 155; Stout *Cooking for Northern Wives,* 34–6.

40 Stewart, 'Notes on the Gannetries of Sula Stack and Sula Sgeir', *British Birds* 31 (1938): 282–94.

41 Ibid., 292.

42 'Gannets as Food: Crofters' Surprise at Protest', *The Times,* 27 August 1938, 8.

43 'Ness and the Guga', *Stornaway Gazette & West Coast Advertiser,* 2 September 1938, 6.

44 Letter from F. Fraser Darling to Mr Maurice, 6 April 1938, BTO James Fisher Papers, BTO Archives RB3. See also his comments in *Natural History of the Highlands and Islands,* 195–7.

45 The matter was also raised in Parliament with the Secretary of State for Scotland, who defended the 'local practice' in reply to a question about the threat to gannets from the hunt: see 'Gannets, Sula Sgeir', Commons Sitting, Scotland, Hansard, 342 cc213-4.

46 Sefton Gordon, 'Hazardous Raids on Sula Sgeir', Letter to the Editor, *The Times*, 30 August 1938, 6; J.S. Grant, 'Gannets as Food', Letter to the Editor, *The Times*, 16 September 1938, 8; F.C.R. Jourdain, 'Gannets as Food', Letter to the Editor, *The Times*, 29 September 1938, 8; Sefton Gordon, 'Gannet Harvest', *The Times*, 15 December 1938, 19.

47 R.M. Lockley, 'Gannet Harvest', Letter to the Editor, *The Times*, 20 December 1938, 10. See also R.M. Lockley, 'Gannets of Sula Sgeir', Letter to the Editor, *The Times*, 21 January 1939, 13.

48 'Gannets – Sula Sgeir', Letter from M. Stewart to J. Fisher, 25 February 1940, James Fisher Papers, BTO Archives RB3.

49 Ibid.

50 Fisher and Vevers, 'The Breeding Distribution, History and Population of the North Atlantic Gannet, Part Two: Changes in the World Numbers of the Gannets in a Century', *Journal of Animal Ecology* 13 (1) (May 1944): 60.

51 Ibid.

52 Letter from Tom Harrisson to James Fisher, 18 April 1944, BTO James Fisher Papers, BTO Archives RB3.

53 Thompson, *Customs in Common*, 9.

54 Murray, *The Guga Hunters*, 221; Thompson, *Customs in Common*, 9. Brian Jackman also described the guga hunt as a way in which the men of Ness 'can keep in touch with their Gaelic roots': see Beatty, *Sula*, 1992, 7.

55 For supportive statements about the hunt and the rightness of the amendment see, 'Guga Taken Off the Menu', *The Times*, 21 December 1954, 3; 'Crofter Eager for Gannet', *The Times*, 1 September 1955, 5; 'The Nessmen and the Guga', *Country Life*, 30 December 1954, 230; 'The Sulisgeir Gannets', *Country Life*, 3 March 1955, 608.

56 Lord Hurcombe, Protection of Birds Bill, House of Lords Debate, 20 May 1965, 266, cc 578–628. See also the letter from P. Brown to P. Beedle, Home Office, in which he, as secretary of the RSPB, suggests a licensing of the hunt by NC: 14 September 1961, RSPB Library, 04-01-00.

57 Barca, Lindon, and Root-Bernstein, 'Environmentalism in the Crosshairs', 195.

58 Murgui, 'When Governments Support Poaching', 127–37; Hirschfeld and Heyd, 'Mortality of Migratory Birds Caused by Hunting in Europe', 47–74; 'General Report', in International Committee for Bird Preservation, British Section, Annual Report for 1953, 3; 'Bird Netting in Belgium', ICBP, British Section, Annual Report for 1954, 4.

59 Hirschfeld and Heyd, 'Mortality of Migratory Birds Caused by Hunting in Europe', 3.
60 Barca, Lindon, and Root-Bernstein, 'Environmentalism in the Crosshairs', 195.
61 Murgui, 'When Governments Support Poaching', 24, 2014, 129–30.
62 Ibid., 130.
63 Birkhead, *Red Canary*, 15.
64 Hirschfeld and Heyd, 'Mortality of Migratory Birds Caused by Hunting in Europe', 3.
65 Capatti and Montanari, *Italian Cuisine*, xv.
66 Ibid., 100.
67 A. Carluccio and P. Carluccio, *Carluccio's Complete Italian Food*, 45–55.
68 Barca, Lindon, and Root-Bernstein, 'Environmentalism in the Crosshairs', 195; Meiklejohn, 'Wild Birds as Human Food', 80–3.
69 'Song Birds as Food', *The Times*, 1 April 1916, 3; Lady Mayo, Letter to the Editor, *The Times*, 31 March 1916, 9.
70 'General Report', in ICBP, British Section, Annual Report for 1953, 3.
71 Bannerman, *The Birds of Cyprus*, xxxvi, 85.
72 'Bird Netting in Belgium', ICBP, British Section, Annual Report for 1954, 4.
73 'Mass Destruction of Birds in Italy', ICBP, British Section, Annual Report, 1958, 6.
74 Durrell, *Bitter Lemons*, 45.
75 'Mass Destruction of Migratory Birds', ICBP British Section, Annual Report 1969, 9–10.
76 See the following reports that appeared in *Birds*: 'Bird Protection in Cyprus', July–August 1965, 326; 'Bird Protection in Italy', November–December 1965, 386; 'Conservation in Italy', September–October 1967, 237.
77 R. Porter, 'Gentille Alouette', *Birds* (March–April 1970): 38–40.
78 D. Lea, 'My Visit to France', 30 October 1972, RSPB Library, Prosecution and law, 04-00-01.
79 'Bird Destruction Abroad', *Birds* (March–April 1970): 49.
80 A.J. Stagg, 'Death on an Island', *Birds* (January–February 1970): 17–18; P. Conder, 'Bird Trapping in Cyprus', Letter to the Editor, *The Times*, 21 April 1973, 13.
81 'Ban on Pickled Songbirds Urged', *Sunday Times*, 22 July 1973, 6; 'Europe's Birds Run Gauntlet of Death', Special Report by Brian Jackman, *Sunday Times*, 26 October 1975, 27–30.
82 'Europe's Birds Run Gauntlet of Death', Special Report by Brian Jackman, *Sunday Times*, 26 October 1975, 27–30.

83 'A European Committee for the Protection of Migratory Birds', A proposal by Dr J.H. Westerman, Chair of Netherlands Committee for Protection of Migratory Birds, 2 March 1976; Letter from Ian Prestt to Robert Adams, WWF, 28 July 1976, ICBP British Section; 'Minutes of the First Meeting of the British Committee for the Prevention of Mass Destruction of Migratory Birds', 4 August 1976; 'RSPB's Involvement with the Problem of the Mass Catching of Birds', Peter Conder, 18 September 1975, all in 'Saving Migratory Birds', Papers of Peter Conder, Bodleian Library, University of Oxford.

84 Woldhek, *Bird Killing in the Mediterranean*, 1980, quoted in M.-A. Falzon, 'Flights of Passion', 15–20

85 Axell, *Of Birds and Men*, 194–5.

86 'EEC Acts to Protect Songbirds', *The Times*, 14 December 1977, 11.

87 Hirschfeld and Heyd, 'Mortality of Migratory Birds Caused by Hunting in Europe', 50.

88 Ibid., 50; Barca, Lindon, and Root-Bernstein, 'Environmentalism in the Crosshairs', 190.

89 Barca, Lindon, and Root-Bernstein, 'Environmentalism in the Crosshairs', 195–204; Murgui, 'When Governments Support Poaching', 131; Falzon, 'Flights of Passion', 15–20.

90 Falzon, 'Flights of Passion', 15.

91 In recent years, as EU laws have tightened the terms within which bird hunting can be legally practised, illegal hunting has emerged as a bigger problem. In Cyprus, killing migrant birds was made illegal in 1974, but the scale of illegal hunting remains high. See, Shorrock, 'Bird Killing in Cyprus', 140–53; Bhattacharya, 'Slaughter of the Songbirds'.

92 Williamson, *The Atlantic Islands*, 133.

CONCLUSION

1 'Advance Release', Department of the Interior Information Service, Fish and Wildlife Service, 12 November 1944, Rachel Carson Papers Box 95, Beinecke Rare Book and Manuscript Library, Yale University.

2 Ibid.

3 Rachel Carson, handwritten notes, Rachel Carson Papers Box 95, Beinecke Rare Book and Manuscript Library.

4 The article was published in November 1945 in the magazine *Coronet*: Letter from Rachel Carson to Martha Lupton, *Coronet*, 28 March 1945; Rachel Carson Papers Box 95, Beinecke Rare Book and Manuscript Library.

5 Rachel Carson, 'Ace of Nature's Aviators', Draft of Proposed Article; Letter from Rachel Carson to Mr Wallace, *Readers' Digest*, 20 November 1944; Letter from Rachel Carson to Martha Lupton, *Coronet*, 28 March 1945; Rachel Carson Papers Box 95, Beinecke Rare Book and Manuscript Library.

6 Cited in 'Nicholson's "The Art of Bird Watching"', *The Auk*, XLIX (1) (January 1932), 119.

7 On Lincoln's idea of 'flyways' see Wilson, *Seeking Refuge*, chapter 3.

8 *State of Nature* (bto.org/our-science/publications/state-nature-report).

9 *RSPB Bird Crime* (www.rspb.org.uk/globalassets/downloads/documents/birds-and-wildlife/crime/birdcrime-report-2019.pdf).

10 'RSPB Risks Widening Rift with Rural Britain, Say BASC', 10 October 2020 (https://basc.org.uk/rspb-announcement-risks-widening-rift-with-rural-britain-says-basc/); 'BASC Urges RSPB to Acknowledge Shooting's Benefits', 28 October 2019 (https://basc.org.uk/basc-urges-rspb-to-acknowledge-shootings-benefits-during-review/); 'RSPB Has Shooting in Its Sights', 6 November 2019 (https://basc.org.uk/rspb-has-shooting-in-its-sights/).

11 The US Fish and Wildlife Service suggested in a survey from the early 2000s that 47.8 million Americans watched birds: see Rosen, *The Life of the Skies*, 3. One million people took part in the RSPB's annual Big Garden Birdwatch in 2021: see 'One Million Take Part in 2021 Big Garden Birdwatch', *British Birds* 114 (May 2021): 251.

12 For preliminary attempts to address the emotional dimension of human–animal relations see, inter alia, Tague, 'The History of Emotional Attachment to Animals'; Guida, 'Surviving Twentieth Century Modernity, Birdsong and Emotions in Britain'. For an anthropological approach, see Milton, *Loving Nature*, 3–4, 55–69. Historians remain divided on what emotions are, how they should be conceptualized, and how their study should inform historical writing. For a fuller discussion of the history of emotions literature, see, inter alia, Plamper, *The History of Emotions*; 'Forum, History of Emotions', *German History* 28 (1) (2010): 67–80; Bar-Haim, *The Maternalists*, 8–9. I am grateful to Shaul Bar-Haim for his insights into the history of emotions literature.

Bibliography

UNPUBLISHED PRIMARY SOURCES

Baker, J.A. (1926–1986), papers, Albert Sloman Library Special Collections, University of Essex.
– A3/1.2–1.3, general, 1966–1988.
– A6.1–A6.52, fan mail.
BBC Written Archives, Caversham.
– R46/364/2, Natural History.
– T31/385/386, Natural History Unit, TV staff 1957–1964.
– T16/40/300/304/306, audience research.
– WE17/54/1, *Life on Earth*, general.
British Trust for Ornithology (BTO) archives, Thetford.
– Common Bird Census.
– Fisher, J., papers.
– Nicholson, E.M., papers.
– Peregrine Enquiry, 1960–1.
Buxton, E.J.M. (1912–1989), papers, Alexander Library Archive and Manuscript Collection, Bodleian Library, University of Oxford.
Carson, R.L., papers, Yale Collection of American Literature, Beinecke Rare Books and Manuscript Library, Yale University.
– general correspondence, 1942–1973.
– series 1 writings, 1921–1975.
Conder, P.J. (1919–1993), papers, Alexander Library Archive and Manuscript Collection, Bodleian Library, University of Oxford.
Council for Nature Archives, Linnaean Society, London.
Hartley, P.H.T. (1909–1994), papers, Alexander Library Archive and Manuscript Collection, Bodleian Library, University of Oxford.
Mass Observation archive, University of Sussex.
– Bird Nesting Directive 1951, MO-30-1-A-E.
– Day Survey Respondent 426, 12 October 1937 and 17 April 1938; daily survey.
– Harrisson, T., Mass Observation report 534, 30 December 1940.

– Respondent 80, 12 February 1937; DSR 630, 12 January 1938.
National Archives, Public Records Office, Kew.
– Nature Conservancy, CAB 124/2628, FT 297; FT 2/177-9, FT 2/198, FT 3/490, FT 3/491, HO 285/162, RSPB Library, archives and publications.
RSPB Minsmere, warden's reports, 1959–1981.

OFFICIAL PUBLICATIONS AND HOUSE OF COMMONS PARLIAMENTARY PAPERS

Board of Trade, *Staggered Holidays, 1962–3, Cmnd 2105.*
– *Leisure in the Countryside, 1965–6, Cmnd 2928.*
Countryside Commission, *The Planning of the Coastline: A Report on a Study of Coastal Preservation and Development in England and Wales* (London, 1970).
– *Coastal Recreation and Holidays* (London, 1969).
Hansard, House of Commons and House of Lords Sittings, volume 187, 201–72; volume 266, 578–628; volume 247, 1,118–49; volume 266, 578–628; volume 342, 213–14; volume 521, 1,483–584; volume 526, 710–44.

NEWSPAPERS, PERIODICALS, AND JOURNALS

The Audubon Magazine, 1952–1980.
The Auk, 1932–1938.
Bird Notes.
Bird Notes and News.
Birds.
Bird Study.
British Birds.
British Trust for Ornithology, annual reports, 1933–1981.
Country Life, 1940–1983.
Essex Bird Watching and Preservation Society, annual reports, 1949–1962.
The Falconer, 1937–1973.
The Field, 1940–1948.
Junior Birdwatcher, 1955–1965.
Kent Ornithological Society, annual report, 1961.
The New Naturalist: A Journal of British Natural History.
Severn Wildfowl Trust, annual reports.
Shooting Times, 1950–1981.
The Times, 1925–1981.

SECONDARY WORKS

Adams, W.M. *Against Extinction: The Story of Conservation*. London and New York: Earthscan, 2004.

Allsop, K. *Adventure Lit Their Star: The Story of an Immigrant Bird*. London: Penguin Books, 1949/1972. (First published 1949.)

Almeroth-Williams, T. *City of Beasts: How Animals Shaped Georgian London*. Manchester: Manchester University Press, 2019.

Attenborough, D. 'Introduction'. In *Peter Scott: Painter and Naturalist*, by E. Huxley. London and Boston: Faber & Faber, 1993.

Automobile Association. *AA Book of the British Countryside*. London: Drive Publications, 1973.

Axell, H. 'Minsmere (Part II), Establishment and Management of an Artificial Brackish Lake with Resting Island ("The Scrape")'. In *Manual of Wetland Management*, 1973.

– *Minsmere: Portrait of a Bird Reserve*. London: Hutchinson, 1977.

– *Of Birds and Men*. London: Bookguild Publishing, 1992.

Baker, J.A. *The Peregrine*. London: Collins, 1967.

– 'On the Essex Coast'. *Birds*, September/October 1971, 427–31.

Baldwin, J.R. 'Sea Bird Fowling in Scotland and Faroe'. *Folk Life* 12, 1 (1974): 60–103.

Bannerman, D.A. *The Birds of Cyprus*. London: Oliver & Boyd, 1958.

– *Birds of the British Isles*, volume 8, 33–4. Edinburgh: Oliver and Boyd, 1959.

Barbellion, W.N.P. *The Journal of a Disappointed Man*. London: Penguin Classics, 2017. (First published 1919.)

Barca, B., A. Lindon, and M. Root-Bernstein. 'Environmentalism in the Crosshairs: Perspectives on Migratory Bird Hunting and Poaching Conflict in Italy'. *Global Ecology and Conservation* 6 (2016): 189–207.

Bar-Haim, S. *The Maternalists: Psychoanalysis, Motherhood and the British Welfare State*. Philadelphia, PA: University of Pennsylvania Press, 2021.

Barrow Jr, M.V. *A Passion for Birds: American Ornithology after Audubon*. Princeton, NJ: Princeton University Press, 1998.

– *Nature's Ghosts: Confronting Extinction from the Age of Jefferson to the Age of Ecology*. London: University of Chicago Press, 2009.

Barth, E.K. 'The Tameness of Some Scandinavian Waders'. *British Birds* 54, 4 (April 1961): 132–5.

Bartie, A., L. Fleming, M. Freemen, T. Hulme, A. Hutton, and P. Readman. 'Historical Pageants and the Medieval Past in Twentieth Century England'. *English Historical Review* 133, 563 (2018): 866–902.

Beatty, J. *Sula: The Seabird Hunters of Lewis.* London: Michael Joseph, 1992.

Beebe, F.L., and H.M. Webster. *North American Falconry and Hunting Hawks.* Denver, CO: North American Falconry and Hunting Hawks, 1964.

Benton, T. *Natural Relations: Ecology, Animal Rights and Social Justice.* London: Verso, 1993.

Berger, J. 'Why Look at Animals?'. In *About Looking.* New York: Pantheon Books, 1980.

Bhattacharya, S. 'Slaughter of the Songbirds'. *Nature* 529 (28 January 2016).

Bircham, P. *A History of Ornithology.* London: Collins, 2007.

Birkhead, T. *Red Canary: The Story of the First Genetically Engineered Animal.* London: Bloomsbury, 2014.

– *The Most Perfect Thing, Inside (And Outside) A Bird's Egg.* London: Bloomsbury, 2016.

Blaine, G. *Falconry.* London: Philip Alan, 1936.

Bocking, S. 'Conserving Nature and Building a Science: British Ecologists and the Origins of the Nature Conservancy'. In *Science and Nature: Essays in the History of Environmental Science,* edited by M. Shortland, 89–114. Oxford: British Society for the History of Science, 1993.

Brace, C. 'A Pleasure Ground for the Noisy Herds? Incompatible Encounters with the Cotswolds and England, 1900–1950'. *Rural History* 11 (2000): 75–94.

Buxton, J. *The Redstart.* London: Collins, 1950.

Campbell, B., and E. Lack, eds. *A Dictionary of Birds.* London: Poyser, 1985.

Cannadine, D. *The Decline and Fall of the British Aristocracy.* New Haven, CT: Yale University Press, 1990.

Capatti, A., and M. Montanari. *Italian Cuisine: A Cultural History.* New York: Columbia University Press, 2003.

Carluccio, A., and P. Carluccio. *Carluccio's Complete Italian Food.* London: Quadrille Publishing Ltd, 1997.

Carson, R. 'My Favorite Recreation'. In *Lost Woods: the Discovered Writings of Rachel Carson,* edited by L. Lear, 12–13. Boston, MA: The Concord Library, 1998. (First published 1922.)

– *Silent Spring.* New York: Houghton Mifflin, 1962.

– 'Road to the Hawks'. In *Lost Woods: The Discovered Writings of Rachel Carson,* edited by L. Lear, 30–2. Boston, MA: The Concord Library, 1998.

Cocker, M. *Our Place: Can We Save Britain's Wildlife Before It Is Too Late?* London: Vintage, 2018.

Cocker, M., and R. Mabey. *Birds Britannica.* London: Chatto & Windus, 2005.
Cocker, M., with D. Tipling. *Birds and People.* London: Jonathan Cape, 2013.
Cohen, S. 'The Empire from the Street: Virginia Woolf, Wembley & Imperial Monuments'. *MFS (Modern Fictional Studies)* 50, 1 (Spring 2004): 85–109.
Cole, E. 'Blown Out: The Science and Enthusiasm of Egg Collecting in the Oologist's Record, 1921–1969'. *Journal of Historical Geography* 51 (2016): 18–28.
Conder, P. 'Conservation and Bird Protection Part Two'. In *The New Wildfowler in the 1970s,* edited by N.M. Sedgwick, P. Whitaker, and J. Harrison, 260–5. London: Barrie and Jenkins, 1970.
– *RSPB Guide to Birdwatching.* London: Macmillan, 1978.
Cramp, S., and P. Conder. 'The Deaths of Birds and Mammals Connected with Toxic Chemicals in the First Half of 1960'. Report Number 1 of the BTO–RSPB Committee on Toxic Chemicals. RSPB, 1961.
Cramp, S., P. Conder, and J.S. Ash. 'The Deaths of Birds and Mammals from Toxic Chemicals, January–June 1961'. Report Number 2 of the BTO–RSPB Committee on Toxic Chemicals (in collaboration with the GRA). RSPB, 1962.
Cullingworth, J.B. 'Planning for Leisure'. *Urban Studies* 1, 1 (1964): 15–16.
Curry, N. *Countryside Recreation, Access and Land Use Planning.* London, 1994.
Davies, G. 'Networks of Nature: Stories of Natural History Film-Making from the BBC'. PhD Thesis, University College London, 1998.
– 'Science, Observation and Entertainment: Competing Visions of Post-war British Natural History Television, 1946–67'. *Cultural Geographies* 7, 4 (2000): 432–59.
– 'The Return of the Avocet'. In *The Return of the Osprey,* by P. Brown and G. Waterston, 163–87. London: Collins, 1962.
Davis, S. '"Britain an Island Again": Nature, the Military and Popular Views of the British Countryside, c. 1930–1965'. PhD Thesis, University of Cambridge, 2010.
– 'Militarised Natural History: Tales of the Avocet's Return to Post-war Britain'. *Studies in History and Philosophy of Biological and Biomedical Sciences* 42 (2011): 226–32.
Dobinson, H.M., and A.J. Richards. 'The Effects of the Severe Winter of 1962/3 on Birds in Britain'. *British Birds* 57, 10 (October 1964): 373–434.

Dobson, J. 'Beyond the Dark Satanic Mills: An Ecocritical Reading of *A Kestrel for a Knave*'. In *Through the Working Class: Ecology and Society Investigated through the Lens of Labour,* edited by S. Critiano, 107–18. Rome/London: Edizioni Ca'Foscari, 2018.

Dower, M. *The Fourth Wave: The Challenge of Leisure.* London, 1965.

Dunlop, T.R. *In the Field Among the Feathered: A History of Birders and Their Guides.* Oxford: Oxford University Press, 2011.

Durrell, L. *Bitter Lemons.* London: Faber & Faber, 1957.

Evans, D. *A History of Nature Conservation in Britain.* London: Routledge, 1997.

Evans, H. ap. *Falconry.* Edinburgh: John Bartholomew, 1973.

Falzon, M.A. 'Flights of Passion: Hunting, Ecology and Politics in Malta and the Mediterranean'. *Anthropology Today* 24, 1 (2008): 15–20.

Fisher, J. *Watching Birds.* London: Penguin Books, 1941.

– *Bird Recognition Volume 1, Sea-birds and Waders.* London: Pelican Books, 1947.

– 'St Kilda: A Natural Experiment'. *New Naturalist* (Summer 1948): 90–108.

– *The Fulmar.* London: Collins, 1952.

– 'Foreword'. In *My Year with the Woodpeckers,* by H. Sielmann. London: Barrie and Rockcliff, 1959.

Fisher, J., and R.M. Lockley. *Sea-birds: An Introduction to the Natural History of the Sea-birds of the North Atlantic.* London: Collins, 1954.

Fisher, J., and H.G. Vevers. 'The Breeding Distribution, History and Population of the North Atlantic Gannet, Part Two: Changes in the World Numbers of the Gannets in a Century'. *Journal of Animal Ecology* 13, 1 (May 1944): 49–62.

Fitzgerald, R.H. *Falconry's Falconer.* Powys: Angle Books, 2013.

Fraser Darling, F. *Natural History of the Highlands and Islands.* London: Collins, 1947.

Fudge, E. 'Two Ethics: Killing Animals in the Past and the Present'. In *Killing Animals,* edited by The Animal Studies Group, 99–119. Urbana, IL/Chicago, IL: University of Illinois Press, 2006.

Garlick, B. 'Osprey Involvements'. PhD Thesis, University of Edinburgh, 2017.

Gill, C., F. Booker, and T. Soper. *The Wreck of the Torrey Canyon.* London: David & Charles, 1967.

Goebel, S. 'Historicism'. Roundtable with B. Bildhauer, S. Goebel, S. Laube, S. Marchand, and A. Svenson. *German History* 34, 4 (2016): 646–71.

Gouyon, J.-B. *BBC Wildlife Documentaries in the Age of Attenborough*. London: Palgrave, 2019.

Griffin, E. *Blood Sport: Hunting in Britain since 1066*. New Haven, CT: Yale University Press, 2007.

Gruffud, P. 'Reach for the Sky: The Air and English Cultural Nationalism'. *Landscape Research* 16, 2 (1991): 19–24.

Guha, R. *Environmentalism: A Global History*. London: Pearson, 1999.

Guida, M. 'Surviving Twentieth Century Modernity: Birdsong and Emotions in Britain'. In *The Routledge Companion to Animal–Human History*, edited by H. Kean and P. Howell, 367–89. London/New York: Routledge, 2019.

Hall, S. 'For Allon White: Metaphors of Transformations'. In *Carnival, Hysteria and Writing*, edited by A. White, 1–25. Oxford: Oxford University Press, 1993.

Hammond, N. 'Conservation and Sister Organisations, the RSPB'. In *Enjoying Ornithology*, edited by R. Hickling, 158–64. London: T.&A.D. Poyser, 1983.

Haraway, D. *Primate Visions, Gender, Race and Nature in the World of Modern Science*. London/New York: Routledge, 1989.

– *When Species Meet*. Minneapolis, MN/London: University of Minnesota Press, 2007.

Harrisson, T., and P. Hollom. 'The Great Crested Grebe Enquiry: Part 1'. *British Birds* 26, 3 (1932): 62–92.

Hines, B. *A Kestrel for a Knave*. London: Penguin, 2000. (First published 1968.)

Hines, R. *No Way But Gentlenesse: A Memoir of How Kes, My Kestrel, Changed My Life*. London: Bloomsbury, 2016.

Hinton, J. *The Mass Observers: A History*. Oxford: Oxford University Press, 2013.

Hirschfeld, A., and A. Heyd. 'Mortality of Migratory Birds Caused by Hunting in Europe: Bag Statistics and Proposals for Conservation of Birds and Animal Welfare'. *Ber Vogelschutz* 42 (2005): 47–74.

Horowitz, D. *Anxieties of Affluence, Critiques of American Consumer Culture*. Boston, MA: University of Massachusetts Press, 2005.

Hosking, E.. *An Eye for a Bird*. London: Readers Union, 1972.

Hosking, E., and S. Smith. *Birds Fighting*. London: Faber & Faber, 1955.

Hoyle, R.W., ed. *Our Hunting Fathers: Field Sports in England after 1850*. London: Carnegie Publishing, 2007.

Huxley, E. *Peter Scott: Painter and Naturalist*. London/Boston, MA: Faber & Faber, 1993.

Huxley, J. *The Courtship Habits of the Great Crested Grebe.* London: Jonathan Cape, 1968. (First published 1914.)

– 'Bird-watching and Biological Science: Some Observations on the Study of Courtship in Birds'. *The Auk* 33, 2 (1916): 142–61.

– 'The Absence of "Courtship" in the Avocet'. *British Birds* 19 (1925): 88–94.

– *Birdwatching and Bird Behavior.* London: Chatto & Windus, 1941. (First published 1930.)

– 'Foreword'. In *Mass Observation,* by C. Madge and T. Harrisson. London: Frederick Muller, 1937.

– 'Birds and Men on St Kilda'. *Geographical Magazine* 10 (1939): 79–80.

– 'Birds and Science'. *Country Life,* 4 August 1950, 367–8.

– 'Introduction'. In *A Field Guide to the Birds of Britain and Europe,* edited by R. Tory Peterson, G. Mountfort, and P.A.D. Hollom. London: Collins, 1954.

– 'Problems of Bird Life'. *Country Life,* 24 June 1954, 2,082–4.

– *Memories.* London: George Allen and Unwin, 1970.

Innes-Smith, R. 'James Wentworth Day (1899–1983)'. *Oxford Dictionary of National Biography.* Oxford: Oxford University Press, 2008.

Johnson, K.R. 'The Ibis, Journal of the British Ornithologists' Union: A Pre-synthesis Portrait'. MSc Thesis, Oregon State University, 2000.

Jones, D. *Birds at My Table: Why We Feed Wild Birds and Why It Matters.* Ithaca, NY/London: Cornell University Press, 2018.

Kallsberg, E. 'The Faeroes Today'. In *The Atlantic Islands: A Study of the Faeroe Life and Scene,* by K. Williamson, 305–21. London: Routledge & Kegan Paul, 1970. (First published 1948.)

Kastner, J. *A World of Watchers.* New York: Alfred A. Knopf, 1986.

Kearton, R., and C. Kearton. *British Birds' Nests: How, Where and When to Find and Identify Them.* London/New York/Toronto/Melbourne: Funk & Wagnalls, 1895.

– *Our Rarer British Breeding Birds, Their Nests, Eggs and Summer Haunts.* New York: Funk & Wagnalls, 1899.

– *British Birds' Nests, Revised and Enlarged Edition.* London: Cassell & Co., 1907.

Kinkela, D. 'The Ecological Landscapes of Jane Jacobs and Rachel Carson'. *American Quarterly* 61, 4 (2009): 905–29.

Lack, D. *The Life of the Robin.* London: F&G Witherby, 1943.

Lambert, R.A. *Contested Mountains: Nature, Development and Environment in the Cairngorms Region of Scotland, 1880–1980.* Cambridge: White Horse Press, 2001.

Landry, D. *The Invention of the Countryside: Hunting, Walking and Ecology in English Literature 1671–1831*. London: Palgrave, 2001.

Larkin, P. 'Going Going'. In *High Windows*. London: Faber & Faber, 1974.

Law, J., and M. Lynch. 'Lists, Field Guides, and the Descriptive Organization of Seeing: Birdwatching as an Exemplary Observational Activity'. *Human Sciences* 11 (1988): 271–303.

Lear, J. *Man and Wildfowl*. London: T.&A.D. Poyser, 1990.

Lear, L.J. 'Rachel Carson's *Silent Spring*'. *Environmental History Review* 17, 2 (1993): 23–48.

– *Rachel Carson: Witness for Nature*. London: Allen Lane, 1997.

Leopold, A. 'A Man's Leisure'. In *Round River: From the Journal of Aldo Leopold*, edited by L.B. Leopold, 3–8. Oxford/New York: Oxford University Press, 1993. (First published 1953.)

Link, F., and M.W. Hornburg. '"He Who Owns the Trifels, Owns the Reich": Nazi Medievalism and the Creation of the Volksgemeinschaft in the Palatinate'. *Central European History* 49 (2016): 208–39.

Lockley, R.M. *I Know an Island*. London: George G. Harrap & Co., 1948. (First published 1938.)

– 'Bird Migration Stations in Britain'. *New Naturalist* (Autumn 1948): 165–71.

– *Letter from Skokholm*. London: Little Toller Books, 2010.

Lowe, P., and J. Goyder. *Environmental Groups in Politics*. London: Routledge, 1983.

Lytle, M.H. *The Gentle Subversive: Rachel Carson, Silent Spring and the Rise of the Environmental Movement*. New York/Oxford: Oxford University Press, 2007.

Mabey, R. *Common Ground: A Place for Nature in Britain's Future?* London: Hutchinson, 1980.

Macdonald, H. '"What Makes You a Scientist is the Way You Look at Things": Ornithology and the Observer 1930–1955'. *Studies in History and Philosophy of Biological and Biomedical Sciences* 33 (2002): 53–77.

– *Falcon*. London: Reaktion Books, 2006.

– *H Is for Hawk*. London: Jonathan Cape, 2014.

Macfarlane, R. 'Introduction'. In *The Peregrine*, by J.A. Baker. New York: New York Review Classics, 2005.

Mackenzie, J.M. *The Empire of Nature: Hunting, Conservation and British Imperialism*. Manchester: Manchester University Press, 1988.

Mandler, P., and S. Pederson. *After the Victorians: Private Conscience and Public Duty in Modern Britain*. London: Routledge, 1994.

Marchington, J. *The History of Wildfowling*. London: A&C Publishers, 1980.

Marren, P. *The New Naturalists*. London: Collins, 2005. (First published in 1995.)

Martin, J. 'Wildfowling and Its Evolution as a Sporting Activity'. In *Our Hunting Fathers: Field Sports in England after 1850*, edited by G. Hoyle. London: Carnegie Publishing, 2007.

Matless, D. 'Visual Culture and Geographical Citizenship in England in the 1940s'. *Journal of Historical Geography* 22, 4 (1996): 424–39.

– 'Moral Geographies of English Landscape'. *Landscape Research* 22, 2 (1997): 141–55.

– *Landscape and Englishness*. London: Reaktion Books, 1998.

– 'Action and Noise Over a Hundred Years: The Making of a Nature Region'. *Body & Society* 6, 3–4 (2000): 141–65.

– 'Versions of Animal–Human: Broadland, c. 1945–70'. In *Animal Spaces, Beastly Places*, edited by C. Philo and C. Wilbert, 115–41. London: Routledge, 2000.

– *In the Nature of Landscape: Cultural Geography on the Norfolk Broads*. London: Wiley International, 2014.

Matless, D., P. Merchant, and C. Watkins. 'Animal Landscapes: Otters and Wildfowl in England 1945–1970'. *Transactions of the Institute of British Geographers* NS 30 (2005): 191–205.

Matless, D., C. Watkins, and P. Merchant. 'Nature Trails: The Production of Instructive Landscapes in Britain, 1960–72'. *Rural History* 21, 1 (2010): 97–131.

Mavrogordato, J. *Behind the Scenes: An Autobiography*. Salisbury: Element Books, 1982.

McKenzie, C.C. 'The Origins of the British Field Sports Society'. *International Journal of the History of Sport* 13, 2 (1996): 177–91.

McKibbin, R. *Classes and Cultures: England 1918–51*. Oxford: Oxford University Press, 1998.

Meiklejohn, M.F.M. 'Wild Birds as Human Food'. *Nutrition and Poultry* 21 (1962): 80–3.

Mellor, G.T. 'Falconry in Britain between 1750 and 1927: The Survival, Organization and Development of the Sport'. PhD Thesis, De Montfort University, 2006.

Milton, K. *Loving Nature: Towards an Ecology of Emotion*. London: Routledge, 2002.

Mitman, G. *Reel Nature: America's Romance with Wildlife on Film*. Washington, DC: University of Washington Press, 2009. (First published 1999.)

Montefiore, J. 'Englands Ancient and Modern: Sylvia Townsend Warner, T.H. White and Fictions of Medieval Englishness'. In *Intermodernism: Literary Culture in Mid-Twentieth Century Britain,* edited by K. Bluemel. Edinburgh: Edinburgh University Press, 2009.

Mort, F. *Capital Affairs: The Making of the Permissive Society.* New Haven, CT/ London: Yale University Press, 2010.

Moss, S. *A Bird in the Bush: A Social History of Birdwatching.* London: Aurum Press, 2013.

Munz, T. '"My Goose Child Martina": The Multiple Uses of Geese in the Writings of Konrad Lorenz'. *Historical Studies in the Natural Sciences* 41, 4 (2011): 405–46.

Murgui, E. 'When Governments Support Poaching: A Review of Illegal Trapping of Thrushes (Turdus spp) in the Paramy of Comunidad Valenciana, Spain'. *Bird Conservation International* 24 (2014): 127–37.

Murphy, R.A. *The Peregrine Falcon.* New York: Houghton Mifflin, 1963.

Murray, D.S. *The Guga Hunters.* Edinburgh: Birlinn Ltd, 2015.

Mynott, J. *Birdscapes: Birds in Our Imagination and Experience.* Princeton, NJ: Princeton University Press, 2009.

Nagai, K., K. Jones, D. Landry, M. Mattfeld, C. Rooney, and C. Sleigh. *Cosmopolitan Animals.* London: Palgrave, 2015.

Nature Conservancy, WAGBI, and Wildfowl Trust. *The Story of a Triumvirate: Wildfowl Conservation in Great Britain.* London: Nature Conservancy, 1970.

Nelson, B. *The Gannet.* London: T.&A.D. Poyser, 1978.

Nethersole-Thompson, D. *The Greenshank.* London: Collins, 1951.

– *The Snow Bunting.* Edinburgh/London: Oliver & Boyd, 1966.

– *Highland Birds.* Edinburgh: Highlands and Islands Development Board, 1971.

– *In Search of Breeding Birds: Early Essays by Desmond Nethersole-Thompson Reprinted from The Oologists' Record.* Leeds: Peregrine Books, 1992.

Nethersole-Thompson, D., and M. Nethersole-Thompson. *Waders: Their Breeding, Haunts and Watchers.* London: A&C Black, 1986.

Newton, I. *Farming and Birds.* London: Collins, 2018.

Nichols, J. *Birds of Marsh and Mere and How to Shoot Them.* London: A.M. Philpott Ltd, 1926.

Nicholson, E.M. *Birds in England: An Account of the State of Our Bird Life and a Criticism of Bird Protection.* London: Chapman & Hall, 1926.

– *The Study of Birds.* London: Benn, 1929.

– *The Art of Bird Watching: A Practical Guide to Field Observation.* London: Witherby, 1931.

– 'Preface'. In *St Kilda Summer*, by K. Williamson and J. Morton Boyd, 11–13. London: Hutchinson, 1960.
– 'Advances in British Nature Conservation'. *SPNR Handbook* (1965), 25–6.
– 'National Wildfowl Refuges'. In *The New Wildfowler in the 1970s*, edited by N.M. Sedgwick, P. Whitaker, and J. Harrison, 216–35. London: Barrie & Jenkins, 1970.
– 'Sir Julian Huxley as an Ornithologist'. *Bird Study* 22, 2 (1975): 113.
– 'Tom Harrisson'. *Ibis* 118 (1976): 596–7.
– 'The Trust – Origins and Early Days'. In *Enjoying Ornithology*, edited by R. Hickling. London: T.&A.D. Poyser, 1983.
Niemann, D. *Birds in a Cage: Warburg POW Camp 1941, Four British Bird Watchers, the Unlikely Beginnings of Modern Wildlife Conservation*. London: Short Books, 2012.
Norrevang, A. 'Traditional Seabird Fowling in the Faeroes: An Ecological Basis for Sustained Fowling'. *Ornis Scandinavia* 17, 3 (1986): 275–81.
Oddie, B. *Bill Oddie's Little Black Bird Book*. London: Methuen, 1980.
Parsons, C. *True to Nature*. Cambridge: Patrick Stephens Ltd, 1982.
Perry, R. *At the Turn of the Tide: A Book of Wild Birds*. London: Lindsay Drummond, 1938.
– 'A Creed for Bird Watching'. *The Field*, 10 February 1940, 214.
– *At the Turn of the Tide: A Book of Wild Birds*. London/New York: Croom Helm, 1973.
Philo, C., and C. Wilbert. *Animal Spaces, Beastly Places: New Geographies of Human–Animal Relations*. London: Psychology Press, 2000.
Plamper, J. *The History of Emotions: An Introduction*. Oxford: Oxford University Press, 2012.
Plummer, K.E., K. Risely, M. Toms, and G. Siriwardena. 'The Composition of British Bird Communities Is Associated with Long-Term Garden Bird Feeding'. *Nature Communications* 10 (2019), article 2088.
Prestt, I. *British Birds, Lifestyles and Habits*. London: Harper Collins, 1982.
Ratcliffe, D.A. 'The Status of the Peregrine in Great Britain'. *Bird Study* 10, 2 (1963): 59–90.
– *In Search of Nature*. Leeds: Peregrine Books, 2000.
Robinson, P. *Falconry in Britain: A Report to the League Against Cruel Sports*. London: League Against Cruel Sports, 1991.
Rosenthal, E.J. *Birdwatcher: The Life of Roger Tory Peterson*. Guilford: Lyons Press, 2008.
Samstag, T. *For the Love of Birds: The Story of the Royal Society for the Protection of Birds 1889–1988*. Sandy: RSPB, 1988.

Savage, M. *Identities and Social Change in Britain since 1940: The Politics of Method.* Oxford: Oxford University Press, 2010.
Schei, P.J. 'Ninety Years and Growing: How Birdlife Went from a Council of Experts to Everybody's Global Partnership'. *Handbook of the Birds of the World.* Madrid/London: Lynx Edicions, 1992.
Scott, P. *Morning Flight: A Book of Wildfowl.* London/New York: Country Life Ltd, 1935.
– *Wild Chorus.* London/New York: Country Life Ltd, 1938.
– *The Eye of the Wind.* London: Hodder & Stoughton, 1961.
Sedgwick, N.M., P. Whitaker, and J. Harrison, eds. *The New Wildfowler.* London: Herbert Jenkins, 1961.
Sharrock, J.T.R. *The Atlas of Breeding Birds in Britain and Ireland.* Tring: BTO/Irish Wildbird Conservancy, 1976.
Sheail, J. *Nature in Trust: A History of Nature Conservation in Britain.* London: John Blackie and Son, 1976.
– *Pesticides and Nature Conservation: The British Experience 1950–1975.* Oxford: Clarendon Press, 1985.
– *Nature Conservation in Britain: The Formative Years.* London: HMSO, 1998.
– *An Environmental History of Twentieth Century Britain.* London: Palgrave, 2002.
Shepherd, N. 'Wild Geese in Glen Callater'. In *Wild Geese: A Collection of Nan Shepherd's Writings.* Cambridge: Galileo Publishers, 2018. (First published 1959.)
Shorrock, G. 'Bird Killing in Cyprus'. *Cyprus Review* 15, 1 (2003): 140–53.
Shrubb, M.F. *Fowling and Feathers: A History of the Exploitation of Wild Birds.* London: T.&A.D. Poyser, 2013.
Sideris, L.H. 'Fact and Fiction, Fear and Wonder: The Legacy of Rachel Carson'. *Soundings* 91, 3–4 (2008): 335–69.
Sielmann, H. *My Year with the Woodpeckers.* London: Barrie and Rockcliff, 1959.
Soper, T. *The Bird Table Book.* London: David & Charles, 1965.
Spencer, R. 'The Trust from 1951 to 1982'. In *Enjoying Ornithology,* edited by R. Hickling. London: T.&A.D. Poyser, 1983.
Spring, I. 'Lost Land of Dreams, Representing St Kilda'. *Cultural Studies* 4, 2 (1991): 156–7.
Stallybrass, P., and A. White. *The Politics and Poetics of Transgression.* London: Methuen, 1986.
Stanford, J.K. *A Bewilderment of Birds.* London: Rupert Hart-Davis, 1954.
Steel, T. *The Life and Death of St Kilda.* London: Harper Press, 2011. (First published 1975.)

Stewart, M. 'Notes on the Gannetries of Sula Stack and Sula Sgeir'. *British Birds* 31 (1938): 282–94.

Stout, M.B. *Cooking for Northern Wives: Containing Practical Recipes for Old-Time and Modern Dishes.* Lerwick: Shetland Heritage Publications, 1925.

Tague, I.H. 'The History of Emotional Attachment to Animals'. In *The Routledge Companion to Animal–Human History,* edited by H. Kean and P. Howell, 345–66. London/New York: Routledge, 2019.

Tansley, A. *Our Heritage of Wild Nature: A Plea for Organised Nature Conservation.* London: The University Press, 1945.

Thomas, K. *Man and the Natural World: Changing Attitudes in England 1500–1800.* London: Allen Lane, 1983.

Thompson, D. 'Contributions to Field Ornithology – In Search of Nesting Birds'. In *Nature's Conscience: The Life and Legacy of Derek Ratcliffe,* edited by D. Thompson, H. Birks, and J. Birks. London: Langford Press, 2015.

Thompson, E.P. *Customs in Common.* London: The Merlin Press, 1991.

Thomson, I.M. *Birds from the Hide.* London: A&C Black, 1933.

Tideman, S., and A. Gosler, eds. *Ethno-ornithology: Birds, Indigenous Peoples, Culture and Society.* London/Washingston, DC: Earthscan, 2010.

Tinbergen, N. *Curious Naturalists.* London: Pan, 1958.

– 'Recent British Contributions to Scientific Ornithology'. *The Ibis* 101 (1959): 130–1.

Toogood, M. 'Modern Observations: New Ornithology and the Science of Ourselves, 1920–1940'. *Journal of Historical Geography* 37 (2011): 348–57.

Tory Peterson, R. *Birds Over America.* New York: Dodd Mead, 1948.

Townsend Warner, S. *T.H. White: A Biography.* Oxford: Oxford University Press, 1989.

Vicedo, M. *The Nature and Nurture of Love: From Imprinting to Attachment in Cold War America.* Chicago, IL/ London: University of Chicago Press, 2009.

Warren, L.S. *The Hunter's Game, Poachers and Conservationists in Twentieth-Century America.* New Haven, CT/London, Yale University Press, 1997.

Wentworth Day, J. *The Modern Fowler: With a Guide to Some of the Principal Coastal Wildfowling Resorts of Today.* London: Batchworth Press, 1949. (First published 1934.)

– *Sporting Adventure.* London: Penguin Books, 1943. (First published 1937.)

– *Wild Wings and Some Footsteps.* London: Blandford Press, 1948.

– *Broadland Adventure.* New York: Charles Scribner's Sons, 1951.

– *The Poison on the Land: The War on Wildlife and Some Remedies.* London: Eyre & Spottiswoode, 1957.

White, T.H. *England Have My Bones*. London: Collins, 1952. (First published 1936.)

– *The Goshawk*. London: Penguin, 1951.

Willcock, C. 'Foreword'. In *Birds of Prey* by P. Brown. Norwich: Survival Books, 1964.

Williamson, K. *The Atlantic Islands: A Study of the Faeroe Life and Scene*. London: Routledge & Kegan Paul, 1970. (First published 1948.)

Wilson, R.M. *Seeking Refuge: Birds and Landscapes on the Pacific Flyway*. Seattle, WA/London: University of Washington Press, 2010.

Woodford, M. *A Manual for Falconry*. London: A&C Black, 1967. (First published 1960.)

Index

Page numbers in italics refer to illustrations.